Science and Representative Democracy

Also available from Bloomsbury:

Advances in Experimental Philosophy of Science, edited by Daniel A. Wilkenfeld and Richard Samuels
Biopolitics After Neuroscience, Jeffrey P. Bishop, M. Therese Lysaught and Andrew A. Michel
Michel Serres and the Crises of the Contemporary, edited by Rick Dolphijn
Michel Serres and French Philosophy of Science, Massimiliano Simons

Science and Representative Democracy

Experts and Citizens

Mauro Dorato

BLOOMSBURY ACADEMIC
LONDON • NEW YORK • OXFORD • NEW DELHI • SYDNEY

BLOOMSBURY ACADEMIC
Bloomsbury Publishing Plc
50 Bedford Square, London, WC1B 3DP, UK
1385 Broadway, New York, NY 10018, USA
29 Earlsfort Terrace, Dublin 2, Ireland

BLOOMSBURY, BLOOMSBURY ACADEMIC and the Diana logo are trademarks of Bloomsbury Publishing Plc

First published in Great Britain 2023
This paperback edition published 2024

Copyright © Mauro Dorato, 2023

English Language Translation © Laura Centonze, 2023

Mauro Dorato has asserted his right under the Copyright, Designs and Patents Act, 1988, to be identified as Author of this work.

For legal purposes the Acknowledgements on p. ix constitute an extension of this copyright page.

Cover image: Social Community with Digital Tablet (© nihatdursun / Getty Images)

All rights reserved. No part of this publication may be reproduced or transmitted in any form or by any means, electronic or mechanical, including photocopying, recording, or any information storage or retrieval system, without prior permission in writing from the publishers.

Bloomsbury Publishing Plc does not have any control over, or responsibility for, any third-party websites referred to or in this book. All internet addresses given in this book were correct at the time of going to press. The author and publisher regret any inconvenience caused if addresses have changed or sites have ceased to exist, but can accept no responsibility for any such changes.

A catalogue record for this book is available from the British Library.

A catalog record for this book is available from the Library of Congress.

ISBN: HB: 978-1-3502-7772-4
PB: 978-1-3502-7776-2
ePDF: 978-1-3502-7773-1
eBook: 978-1-3502-7774-8

Typeset by RefineCatch Limited, Bungay, Suffolk

To find out more about our authors and books visit www.bloomsbury.com and sign up for our newsletters.

To my family

Contents

List of Figures		viii
Acknowledgments		ix
Introduction		1
1	Historical Prologue: The Lippmann–Dewey Debate	9
2	How Does Science Work? The Evaluation and Controllability of Scientific Hypotheses	25
3	How Does Democracy Work? The Balance of Powers	47
4	Representative Democracy, Direct Democracy, and Scientific Specialization	63
5	Scientific Disinformation and Distrust in the Experts	81
6	How to Navigate in the Disagreement of Experts: The Need for Greater Scientific Literacy	105
7	The Role of the History and Philosophy of Science in the Democratic Debate	125
Conclusion		155
Notes		157
References		167
Index		179

Figures

2.1	The complete spectrum of electromagnetic radiation	27
2.2	The phenomenon of diffraction, describing the passage of a light wave through two slits of different aperture	31
2.3	The spatiotemporal curvature explains the motion of the gravitating objects	32
5.1	The Moon as painted by Galileo in his *Sidereus Nuncius*, published in 1610	86
7.1	Each node represents the belief of a community of citizens, or of a group of scientists	143

Acknowledgments

I thank Raffaello Cortina, editor of Raffaello Cortina Editore, for having transferred me the copyrights of a great part of the book *Disinformazione scientifica e democrazia. La competenza dell'esperto e l'autonomia del Cittadino*, originally published in Italian by Cortina Editore, Milan, 2019. I also thank Laura Maccagni for her help with the copyright of the pictures present in the italian version of the book and Dr Roza I.M. El-Eini who copiedited the manuscript with great care.

The writing of the book in Italian and its translation was supported by the Italian Ministry of Education, University and Research through the PRIN 2017 program, "The Manifest Image and the Scientific Image," prot. 2017ZNWW7F_004. I thank the anonymous referees and my colleague Professor Rosa Calcaterra for their suggestions on a previous draft of the manuscript.

Introduction

The main aim of this book is to investigate the relationship between science and democracy from the perspective of the values and the norms that ground them. Pluralism, with its complex relationship with values like consensus and objectivity, tolerance, respect for facts, impartiality, and openness to criticism and doubt, has played and still plays an important role both in the growth of scientific knowledge and in the rules governing a democracy. Conformism and dogmatism have been and still are obstacles to the well-functioning of both science and democracy. Is there a common reason that explains the shared commitment to these values?

Even considering the undeniable differences in the decision-making mechanisms that science and democracy use to reach their respective objectives, my fundamental methodological premise is that *problem-solving* represents the essential aim of both scientific knowledge and democratic institutions, with the term "problem" being considered in its broadest possible meaning.[1] Two very important consequences deriving from this fundamental premise (henceforth, T_1 and T_2) will be the subject of a thorough investigation since they are the fundamental theses around which this book is shaped. They are so strictly intertwined that they can be regarded as the two faces of the same coin, given that they depend in their turn on two highly interdependent facts (F_1 and F_2) that, as such, cannot be questioned.

T_1. Given the progressive specialization of scientific knowledge, which is the undeniable factual consideration F_1, the **first thesis** argues that such a specialization makes *representative* and *indirect* kinds of democracy preferable as well as unavoidable. As we shall see, the most important reason why the former kind of political systems must be preferred to their populistic rivals—based on the utopian ideals of direct democracies—significantly depends on the risks of new forms of communication allowed by the web, which generates consensus without any mediation by experts.

T_2. The **second thesis** is that a well-functioning democracy must be based on the highest possible level of scientific literacy.[2] In order to clarify at the outset my reading of T_2, I should specify that for my purposes it will not prove necessary to inquire in depth into the important but well-explored issue of the so-called "public understanding of science." To the task of placing T_2 in the extensive literature devoted to this issue, two sections in Chapter 6 (6.2 and 6.3) will be sufficient.

The fact, F_2, that justifies this thesis is given by the increasing role that technology, and not only information technology, plays within contemporary Western societies. I shall argue that without some familiarity with the history and philosophy of science as well as some key notions of scientific methodology, scientific disinformation traveling at the speed of light on the web (Temming 2018a–b) can put at risk the life of thousands of human beings by spreading, for instance, false beliefs about dangerous medical practices. Conversely, therapies or vaccines that are indispensable to our health are rejected due to sheer ignorance. The fact that climate change, caused by human beings, can make life on Earth impossible in a few years,[3] is denied even by some scientists.[4]

However, there is an implicit conflict between the ideals grounding representative democracies and the principle of competence that must be made explicit since it will be a *leitmotiv* of this book. On the one hand, at least in principle,[5] it is only in representative democracies that, thanks to free elections, it is possible to choose delegates that are more competent than us and therefore more capable of finding the best means to realize our objectives, aiming, as much as possible, for the public interest. On the other hand, because of the progressively more rapid processes of scientific specialization, there is the risk that liberal democracies turn into *technocracies*, that is, regimes where a very small number of experts decides for all citizens.

To prepare the ground for the discussions of these two theses (in particular T_1), in Chapter 1, I will revisit a debate between the journalist and essayist Walter Lippmann (1889–1974) and the philosopher and educational theorist John Dewey (1859–1952) that took place in the first two decades of the twentieth century and that concerned the conflict between the experts' competence and the citizens' autonomy of political choice. The importance of recovering this exchange in the present context is motivated by the fact that my evaluation of some of their claims will be put to use in the subsequent chapters. Given the theoretical character of this volume, however, there will be no attempt to give an historical reconstruction through to the current time of the issues they debated one hundred years ago.

In Chapter 2, I shall synthetically but as accurately as possible answer the ambitious question: "How does science work?," by stressing the role that the public controllability of a hypothesis plays in the formation of *consensus* in a scientific community. Even if this controllability is realized in different ways in different sciences, (i) the openness to criticism and doubts, (ii) the competition between rival research groups, (iii) the trust in the beliefs of scientists whose competence is widely recognized, (iv) the importance for scientists not to lose their reputation by defending a theory that they know is the product of a fraud, and (v) the process of selection of the results of scientific research in peer-reviewed journals, conferences, and specialized books, are characteristic of *all* empirical sciences.

Science works because the above-mentioned principles facilitate the elimination of errors. If accepting a hypothesis as well confirmed from a scientific viewpoint entails a collective decisional process grounded in a critical evaluation based on the competence of the experts, we shall see that—by considering the relevant differences—also the decisional processes of a well-functioning democracy are based (and should be based) on public discussions, openness to criticism from the opposition, and a free press. Focusing on a somewhat idealized form of democracy is justified by the assumption that guides this book and that is stated at its beginning: my approach is neither descriptive nor sociological, but is rather *normative*, centered on the values that lie at the foundations of science and democracy, the two great institutions of the modern world, which can be regarded as born together.

In Chapter 3, focusing in more detail on these issues, I shall raise a similar question about democracy: "How should an ideal democracy work?" With inevitable simplifications, but in a sufficiently accurate way, I will briefly discuss three essential characteristics of representative democracy: the *equality principle*, the *majority principle* in the realization of the idea of the *government of people*, and the principle of the *separation of powers*. Each of these principles has deep analogies and instructive differences from those ruling and enabling scientific knowledge, in particular those concerning the origin of consensus and the causes of dissent.

The models of representative and direct, populist democracy that I shall discuss in Chapter 4 are an articulation of the concept of democracy, despite the important differences that the two models assign to the above-mentioned three principles. The defense of a representative democracy to be discussed in this chapter (i.e., the first thesis T_1 mentioned above) is essentially based on the "*division of cognitive or epistemic labor*" that Philip Kitcher has insisted upon in many of his publications.[6] In a nutshell, this is the fact that knowledge is not equally distributed among citizens and even among experts specialized in different fields.

The beliefs that we can directly verify with our individual experience are very few: the epistemic reliability and solidity of most of our beliefs, scientific or not, depend on the reliability and solidity of other people's beliefs: almost all of our beliefs have a social origin. No theory of progress of human knowledge—and the increased possibilities to modify the environment to our advantage—can be advanced without considering its essential social character, based on the plasticity of our brain and therefore on our capacity to learn from others. However, the more scientific knowledge grows, the more it specializes, and the more it becomes inaccessible to the public at large.

Consequently, and understandably, citizens feel perplexed and may develop a negative attitude toward scientific expertise and the principle of competence in general. On the one hand, our necessary dependence on the experts' beliefs and knowledge limits the autonomy of their decisions. On the other, in the numerous cases of *conflicts of opinion* among experts or scientists, possibly motivated by economic interests,[7] citizens do not know whom to trust. Such conflicts of opinion easily generate in uninformed citizens the conviction that no hypothesis or belief is well founded.

The consequent, widespread suspicion toward the "*mediators*" of knowledge (i.e., the experts) generate, together with other factors, distrust of the mechanisms of political mediations and representative institutions. This is one of the reasons that explains the contemporary appeal to different forms of *populism*, accompanied by the temptation of adopting direct kinds of democracy, which apparently allow a greater autonomy of citizens' decisions. One of the main causes of populism is the illegitimate classification of *all* scientists and experts as an elite on the payroll of some powerful companies or of a few powerful financial groups. These groups are often regarded as being responsible for the increasing and disquieting gap between the rich and poor, as well as for the progressive impoverishment and fading of the middle class, which is the basis of liberal democracies. Even though a careful analysis of these important socioeconomic factors cannot be undertaken here, it seems plausible to suppose that the cultural lack of scientific literacy and the social weakening of the middle class are among the main causes contributing to the spread of fake news.

Proponents of direct democracy can appeal to contemporary technology, thanks to which citizens would be *free* to push a button on the screen of a smartphone, to answer with a "like" or "dislike" questions that are accurately chosen and posed by the small number of people who are really holding the power. Paradoxically, *direct* democracies have very good chances to become *heterodirect*, that is, directed by a few others. Not only do the concepts of *dis*information and *mis*information presuppose that (1) there is correct and objective scientific information that corresponds to real facts but also that, in the relevant cases, (2) it is possible to distinguish objectively beliefs that are defended by real experts from those advertised by self-proclaimed ones.

For this reason, in Chapter 4, I will show why the thesis that all scientists are influenced by their political opinions and by the social context in which they live and work does *not* entail that their hypotheses and beliefs are not objective and impartial and that they are valid only within that specific context. To defend this point of view, it is important to distinguish between two senses of *objective*: that is, (i) *intersubjectively valid*, and (ii) *independent of the human mind* or *mind-independent*. I will then move to criticize some arguments, usually based on the historical development of science, and aimed at denying the possibility of achieving objective knowledge (in the two senses of the word). In the central part of the chapter, I will discuss three kinds of disagreement among real experts on the one hand and charlatans or *merchants of smoke and doubts*[8] on the other hand. In the final part of the chapter, I will examine the problem of the neutrality of science (the relationship between science and values) in those cases in which true hypotheses are considered false and conversely (the so-called *inductive risk*).

In Chapter 5, I will provide additional evidence in favor of the second thesis (T_2) by showing why the survival of liberal democratic institutions presupposes an increased scientific literacy and knowledge of the key notions in scientific methodology which cuts across and unifies the various scientific disciplines. There are at least two strictly related reasons in favor of this claim. First, there exists a latent conflict between the increasingly important role that high-tech devises play in Western societies and the fact that citizens must decide as autonomously as possible on ethical issues generated by exponentially accelerating technological changes. Second, due to the growing experts' specialization, citizens cannot possess universal knowledge, a fact already recognized by Lippmann and Dewey (see Chapter 1). Consequently, two important problems that need to be tackled are, (i) how to decide which experts we should trust in case of disagreement between them, and (ii) how to recognize true experts from alleged ones, who can help themselves with mechanisms of propaganda and misinformation which, being spread via the social networks, are obviously faster and more capillary than those used by the fascist regimes in the 1930s (radios and movies).

In Chapter 6, I will strengthen T_1 by presenting the so-called *jury theorem* proposed by the French philosopher and mathematician Nicolas de Condorcet (1743–94). Not only has this theorem the advantage of being explainable in a non-technical language, but it can also be reformulated to show that the political decisions made by a group of citizens *whose beliefs are closer to being true* become more and more reliable the more citizens are added to the original group. The opposite holds for jurors/citizens whose information is closer to being false: in this case, the increase in the number of jurors makes the probability that the final verdict (political decision) is wrong become a certainty. It is appropriate to stress that—among the other premises discussed below—the correctness of the jurors' beliefs plays a key role.

In the concluding Chapter 7, I will discuss the important role that the history, philosophy, and the methodology of science—in particular the *probability* of a hypothesis, its *evidence*, and *causality*—can play to help citizens to autonomously identify pseudoscientific hypotheses and fake news. In order to acquire a scientific attitude, I will claim that the important role that the principle of competence has with respect to democratic decisions does not only concern science, but also the humanities. A greater level of literacy in general therefore entails a more complete perspective on human culture and the values of civilization. Since the actual knowledge of the mechanisms that are at the basis of a scientific hypothesis is an important part of the history and philosophy of science, I conclude by showing in which sense the latter disciplines are the more robust bridges capable of connecting the sciences to the humanities.

1

Historical Prologue

The Dewey–Lippmann Debate[1]

1.1 Introduction

In order to introduce and begin discussing my main claims T_1 and T_2, there is no better choice than focusing on Walter Lippmann's ([1922] 1997, [1925] 1993) and John Dewey's (1927) "debate."[2] There are at least three reasons that motivate this choice:

1. To my knowledge, they posed for the first time clearly and explicitly the main problem that I want to tackle. Both authors shared the view that the growing conflictual nature of the experts/citizens relation was caused by the fact that omniscient citizens could no longer exist.
2. Both were acutely aware of the consequent problem of reconciling the necessity of scientific expertise with a participatory democracy where citizens are given the possibility of autonomous choices. The problem is originated by the fact that even though the two principles (of expertise and autonomous choice) are in potential conflict, ideally they ought to be implemented *in the same institution* if this institution is called to serve our common interest. For simplicity, from now onward, I will refer to this problem with the label **RP** (reconciliation problem).
3. Even if it occurred approximately 100 years ago, their debate will help me to shed light on some of my arguments. In my view, the reason that explains why their dispute is of contemporary interest[3] is that the relevant characteristics of the (few) early-twentieth-century representative democracies are also present in a much more radical way today. If some solutions they proposed to solve the RP could not work in their time, when science was much less

specialized and democratic institutions less complex, *a fortiori* they could not work today. However, it could be the case that in our time, other solutions that the protagonists of the debate deemed impossible could be more easily implemented.

The relevance for our theme of Dewey's interest for Lippmann's work can hardly be exaggerated since the two authors gave two radically different answers to RP: as we will see, they can be regarded, respectively, as the two champions of technocratic and participatory democracies. Dewey claimed that radical reforms in our educational systems striving for the creation of "a society of knowledge" could avoid *both* Lippmann's skepticism about the possibility that citizens choose their representatives without being moved only by propaganda ways *and* over-utopic solutions to RP given by institutions, in which everyone can decide about everything without the mediation of experts.

In the next section, 1.2, I will begin by describing Lippmann's claim about the relationship between experts and the political class on the one hand, and what he defined as "the Phantom Public" on the other.[4] In section 1.3, I will illustrate Dewey's objections to Lippmann by defending his claim that a democratic institution can give a solution to RP *only* on the condition that it can be converted into a "society of knowledge." In such a society, experts would of course keep on having an indispensable role, but thanks to the formation of a "community of inquirers," the entire society and each single component could make more informed and autonomous choices. In my reading of this ideal but not utterly utopian society, in later chapters I will show why—in order to aim for Dewey's model of society—not only should each citizen be acquainted with the social nature of the acquisition and control of scientific hypotheses (Chapter 2) but should also be acquainted, more generally, with the historical and philosophical foundations of science in the sense to be explained in Chapter 7. Traditional forms of representative democracies (with the necessary adjustments that do not destroy its nature) seem the best way both to, (i) strengthen as much as possible the needed interaction between the citizens, the experts, and the politicians, and (ii) to provide the citizens with the capacity to take a reasoned stand in the growing cases in which the experts' opinions about decisions that must be taken urgently *diverge*.

In what follows, it is of fundamental importance to keep in mind that the forms of democracy that I discuss in this and in the following chapters—namely direct and representative democracy—are *ideal types* (*Idealtypen*) in the sense introduced in the literature by the German sociologist and historian Max Weber (1864–1920). An ideal type (*Idealtypus*) is a *model* or an *idealized* pattern by means of which we can understand in a clearer way the essential features of *real* social phenomena: "an ideal type is formed by the one-sided accentuation of one or more points of view," according to which "concrete individual phenomena ... are arranged into a unified analytical construct" (*Gedankenbild*); in its purely fictional nature, it is a methodological "utopia [that] cannot be found empirically anywhere in reality" (Weber 1949: 90). In this sense, the models of direct and representative democracies now to be introduced can be regarded as "ideal types" that contain normative components that partially overlap. For instance, in *all* forms of democracy, experts and elected politicians must be as responsive as possible to the citizens' problems and needs.

In later chapters, when I will present the many conceptual links between the values of science and those of democracy, I will study in much more detail the relationship between direct and representative democracies. In this chapter, as an introduction to the Lippman–Dewey debate, it will suffice to note that the defenders of "ideal-typical," *direct* democracies as I understand them, disparage many forms of merely procedural types of democracy, typically based on political delegation. In these types of institutions, they argue, the politicians are unfaithful to the electoral mandate and often advantage few dominant minorities that never respond to the most pressing needs of the majority of electors.

In addition, the defenders of direct forms of democracy often claim that instruments like referenda, together with new forms of participation granted by the internet and other social networks, would give "the people" much more influence over society and guarantee them more autonomous choices.[5] The defenders of direct democracies add to the criticism of the merely procedural aspects of representative democracies (among which, first and foremost, free elections based on universal suffrage) attempts to empower the citizens in a concrete way by making possible their active intervention in the public sphere via autonomous decisions that involve directly their life. Instruments that ought to be (and are) adopted also by representative democracies usually involve citizens' demonstrations, public assemblies, and other initiatives, thanks to which ethical issues like the legalization of abortion, divorce, matters concerning the end of life, or biological wills can be publicly debated. On the basis of its different and pluralistic ethical orientations, the whole population can then be called to decide via instruments like referenda, thereby avoiding autocratic or technocratic institutions that transform a particular ethical or religious point of view into a state law.

While representative democracies do (and ought to) make room for referenda of this kind, they differ from direct forms of democracies because the former stress more than the latter the fact that there are public issues on which public referenda or polls are ineffective in solving the citizens' problems and can even be dangerous for the society. Think of decisions to use nuclear energy, incinerators for garbage, the necessity to curb or eliminate the production of certain substances that endanger the survival of our planet. Additional, well-known cases, include the necessity of vaccination against certain diseases (measles, mumps, rubella, Covid-19, etc.) and consequent measures of public protection (lockdown vs. herd immunity). In these cases, the public needs first to be thoroughly informed and instructed by experts or competent scientists, and referenda might not be the best instruments to decide how to promote the common good and the common interest.

Therefore, one of the most important differences between representative and participatory democracies essentially relates to the different ways in which *knowledge is transmitted and distributed in scientific communities* and, therefore, to the values that ground the decisional procedures of science in comparison to those characterizing direct democracies. It is not by chance that Dewey argued that the true empowerment of people can be achieved only thanks to a more widespread scientific attitude.

1.2 Lippmann's "Phantom Public"

Lippmann regarded forms of democracies requiring this sort of more technical information as unrealizable for three reasons: (a) the non-existence of omnicompetent citizens; (b) the fact that the vast majority of citizens is subject to propaganda; and (c) that no institution can be imagined that could bridge the gap between the experts giving advice to the politicians and the public.

As to (a), Lippmann is correct in arguing that the ever-growing specialization makes participatory democracies extremely difficult to realize. This claim does not depend so much on the charge that they are utopian but rather on the fact that, lacking some important specifications, they fall into a dilemma from which it is difficult to escape. They *either* slide toward populism, where "the people" is considered as a homogenous mass apparently moved by the same interests but often guided by few autocrats *or* reduce to inane attempts to find difficult mediations among citizens animated by conflicting values—Weber's "polytheism of values" (1958: 148–9)—that cannot be composed and that require some form of technocracy.

As to (b), the problem according to Lippmann is that the public's representation of social reality is not only limited but also systematically *distorted by propaganda*. The public is ignorant and gullible, writes Lippmann, since its opinion is not based on "direct and certain knowledge, but on images that he forms or that are given to it" ([1925] 1993: 74). Lippmann concludes that, given the impossibility of omnicompetent citizens, there must be a group of (independent?) experts whose task is to counsel the decision-makers by supplying them with the information that is necessary to reach a decision. The information cannot be transmitted and elaborated by citizens.

As to (c), in Lippmann's view of the relation between experts and politicians on the one hand and citizens on the other is based a radical distinction between two components of the society, namely agents and spectators. More in detail, according to DeCesare (2012), we must further distinguish the first group (the "agents"), in two subgroups, namely the real decision-makers (administrators, politicians, CEOs, etc.), and the experts (the second "actors"), whose task is to instruct the former on how to reach goals that are not really "chosen" by the citizens, the bystanders, who are (and are bound to remain) ignorant about concrete solutions to social questions that involve them directly.[6] According to Lippmann, (c) is an unavoidable consequence of (a); therefore RP is not solvable by an attempt to create some more participatory structures that, by giving citizens more autonomy of choice, may enable them to intervene *directly* by influencing the politicians' decisions.

For what concerns (a), it is important to note that Dewey *agreed* with Lippmann on the fact that the ideal of the "omnicompetent citizen" is unrealizable and that, as Kitcher put it today,[7] the division of cognitive labor is unavoidable. Evidence for this claim is the following quotation mentioning Lippmann's work "*Phantom Public*": "I wish to acknowledge my indebtedness, not only as to this particular point [the bewilderedness of the public] but for ideas involved in my entire discussion even when it reaches conclusions diverging from his" (Dewey 1927: 116–17). In essence, Dewey shares Lippmann's criticism of democracy as it was structured in their times: in his review in 1922 of Lippmann's *Public Opinion* (Lippmann 1922), Dewey wrote that the book was "the most effective indictment of democracy as currently conceived ever penned" ([1922] 1976: 337).[8]

It follows that the important point of divergence between Lippmann and Dewey is not about (a) or (b), but about (c), since it consists in the degree of optimism with which the latter envisages a solution to RP. To put it differently, Lippmann and Dewey agree on the starting point, namely the contemporary passive role of the public, but they differ on the power that education can have to reinforce the naturally social attitude of human beings. According to Dewey, an education to a scientific way of thinking does not have the dire consequences often envisaged, since it neither engenders a technocratic regime that oppresses the freedom of choice of the citizens nor implies the suffocation of the citizens' autonomy caused by purely procedural institutions.

As to (c), in particular, the difficult question that Dewey must face is how to concretely and non-idealistically bridge the gap between the experts and the public. Since this problem calls into play his whole philosophy of education, it cannot be treated here. Among the various proposals designed to fill the gap between experts and citizens, I can just mention here the possibility of creating councils or citizen's committees, presumably elected by scientific societies. These institutions might enable citizens to intervene more rationally in decisions whose consequences affect their life (Kitcher 2011).

In sum, the solution that Lippmann offers to RP is certainly closer to technocracy as I defined it but despite its liberal elitism, it is certainly *not* antidemocratic. As Whipple (2005) claimed, Lippmann is unjustifiably skeptic about the possibility of a better education for the population at large, while Dewey wrote extensively about education as key to a well-functioning democracy (see, for instance, Dewey 1916). According to Lippmann, as for Tocqueville,[9] citizens are either too busy or cannot have access to the truths of science. Consequently, they must be guided by "a centralized body of experts to act as society's intelligence," where the intelligence in question is necessary to the politicians to make "informed, rational decisions" (Whipple 2005: 160). As DeCesare put it, "the average citizen cannot come close to having the scope and depth of undistorted knowledge of the world necessary to manage political affairs" (2012: 109).

However, there are a few problems raised by Lippmann's proposal that need to be discussed and that in his time were not clearly delineated. One of the main difficulties is generated by the fact that today, different *pools* of experts do not speak the same language, so that what Lippmann referred to as "a single pool" is now fragmented into countless others. Consequently, what Lippmann called "pools of experts" contain members who might not be able to understand the language of other members of the *same* pool. In other words, Lippmann could not imagine that many *subgroups* of a single pool might not be able to fully understand each other. Once the pools and the "subpools" of experts become too numerous, their interaction with the politicians becomes quite complicated.

Lippmann is correct in arguing that in representative forms of democracies, the role of the politicians cannot be bypassed, given that they are themselves more *expert* than the average citizen on, say, financial, economical, and juridical problems that they must solve. It is at this crucial juncture that the question of *trust* between politicians, citizens, and experts emerges, and with it the social character of science that, as we will see in the next chapter, is key to its solution: the social character of human beings grounds, but also depends on, the way knowledge is formed and transmitted.

Lippmann's stress on the need for expertise is understandable. However, due to the growing division of cognitive labor, the "epistemic distance" between a citizen C from the experts E (what we could call "C–E distance") *is not superior* to that (1) separating E from experts E' working in different fields (the E'–E distance) and to that (2) separating C from the politicians P (C–P distance). This sort of "equal-distance principle," which is predicated upon a widespread level of scientific literacy, can be justified only if *C, E, E' and P* are aware of the social nature of science, a factor that Lippmann, in his rigid three-partitions of roles, neglected. An awareness of this factor seems the only way to overcome the distance between the experts on the one hand, and the politicians and therefore the voters on the other.

Lippmann missed this social aspect of knowledge not only because of his "atomistic" anthropology (Dewey 1927),[10] but also because, unlike Dewey, he did not have a clear understanding of the methods that make the process of gathering evidence for a scientific theory possible.

As we are about to see in more detail in the next chapter, the social character of science—of which Dewey was fully aware—implies that science is based on the construction of extremely complicated social networks of theories that must form *consistent structures*. Ties of the net are formed by scientists or groups of experts that have a mutual trust and rely on the method of testing in general scientific hypotheses. Albert Einstein (1879–1955) wrote that "scientific concepts are free inventions of the human mind" (Einstein [1933] 1954: 270) since they cannot be simply deduced from the experimental data. We ought to add that "they are free inventions of socially interacting human minds," even though the fact that such theories are socially constructed and fallible does *not* entail, as has often been maintained, that they cannot be approximately true or at least highly confirmed. In this sense, according to Dewey, science is the best form of knowledge.

1.3 Dewey's Criticism of Lippmann: The Methodological Component

Dewey's twofold criticism of Lippmann has a methodological as well as a political-normative component. The former has become a pivotal element in the current debate on scientific realism. The latter, despite its idealistic nature, can be in part justified by his philosophy of education and his attention to an *evolutionary conception of human conduct*: according to Dewey, education was inseparable from democracy, and the former essentially included learning to apply science's experimental method to the most pressing social problems.

From a methodological viewpoint, Dewey's main objection to Lippmann's views was based on his conviction that one's general view of human beings, that is, one's *social anthropology*, must reflect one's *theory of knowledge* (and conversely). In Dewey's view, human beings are highly social animals, and are not like "heaps of grains of sand." Correspondingly, in his epistemology, he stressed a persistent dichotomy between, on the one hand merely *passive* and *representational* views of human knowledge and science (what he called "the spectatorial theory of knowledge"), and on the other *active* and *interventionist* views. Given the inseparability of epistemology and theories of democracies, according to Dewey this dichotomy was reflected in the clash between *passive* and *active* views of the public's role vis-à-vis experts and decision-makers. This, distinction, which essentially characterized Lippmann's political views, for Dewey could be overcome in favor of participatory view of citizenship, in the sense in which science overthrew scholastic philosophy. This aspect of Dewey's philosophy is the strongest argument against Lippmann's elitism.

We can convince ourselves of the strict link between epistemology and political institutions by recalling Aristoteles' metaphysics, in which a merely contemplative view of knowledge was mirrored by the "superior" role of God,[11] *who contemplates himself without acting*. God is changeless because he is pure actuality; as such, he is not subject to the passage from potency to actuality, that is, to time. To the extent that this conception of the supreme being reflects the social structure of the Greek poleis, a contemplative view of human knowledge entailed that the manual work of artisans and peasants was considered to be inferior. As such, it was belittled by the Aristocrats, who did not have to work to earn a living. Despite the fact that this interesting link proposed by Dewey here cannot be further explored, I would like to suggest that Dewey's idea that different conceptions of how human beings acquire knowledge are reflected and reflect different conceptions of social institutions appears extremely plausible, since it is *more* than an analogy.

This epistemological stance is by far Dewey's most convincing argument in favor of the ideal of democracy that he defends. In science, decisions about which hypothesis must be adopted and therefore when further inquiry should be temporarily stopped, are taken after a collective debate engaging all participants, including those who endorse different viewpoints.[12] Dewey correctly noted the indubitable fact that since the scientific revolution, beliefs and empirical hypotheses about the natural world have been based more and more on humans' active interventions (i.e., carefully planned *experiments*) than on passive individual contemplations of the external world that prevailed in *most* Greek science and philosophy.[13] It is not a coincidence that the birth of modern science was favored by a new emphasis on experiments—*a form of interventions due to manual work*—and that Galileo claimed to have learned more from the hand-on work of artisans who had to repair boats in Venetian shipboards than from many philosophical books written on papers. Moreover— going back to the methodological point concerning scientific realism that was mentioned at the beginning of this section—stressing the role of experiments in contemporary science yields the most powerful argument in favor of the existence of entities that are not directly observable. On the contrary, the merely *representative* roles of scientific models must be interpreted as instrumental for successful predictions, given that the latter are not construed to "mirror" nature (Rorty 1979). In a slogan, science is more a matter of "intervening rather than representing,"[14] and it is not coincidental that Hacking, the authoritative philosopher of science that stressed more than anybody else this view of the philosophy of science, explicitly referred to Dewey in many passages of his book (Hacking 1983: 61, 62, 63).[15]

In Dewey's thought, science coincides with an experimental attitude directed to problem-solving, since: "The growth of democracy is connected to the development of the experimental method in science, to evolutionary ideas in biology and to the industrial reorganization" (1916: 3). In this respect, Dewey's literacy in the natural sciences and his deeper understanding of the scientific method, conjoined with his familiarity with Darwin's evolutionism (Dewey [1907–9] 1977), gave him an important advantage over a learned and brilliant journalist who, however, was not familiar with science. In this respect, MacGilvray, by directly quoting Dewey, has reminded us that for him, "while the rise of modern science has revolutionized our understanding of the material world, our approach to moral questions remains mired in traditional, pre-scientific ways of thinking" (quoted in MacGilvray 2010: 36). Passages like these show more than others why for Dewey, Darwin's evolutionary theory was a fundamental component to understand the origins of morality.

1.4 Dewey's Criticism of Lippmann

Dewey's criticism of Lippmann, however, is more controversial in its normative-political part.[16] His starting point is the reasonable assertion that an ideal democracy should not coincide "with a government by experts in which the masses do not have the chance to inform the experts as to their needs." This would amount to "an oligarchy managed in the interests of the few" (Dewey 1927: 225). He and Lippmann agreed that this was a necessary condition for an institution to be a democracy.[17] In a democracy, there must be feedback between the citizens and the expert and with a well-known, very telling metaphor, Dewey writes that: "The man who wears the shoe [the citizen] knows best that it pinches and where it pinches, even if the expert shoemaker is the best judge of how the trouble is to be remedied" (1927: 217). Given that this ideal is essential to democracy, *how* can we realize it? As hinted above, Lippmann was convinced that it could never be achieved.

On this point, however, Dewey's answer to Lippmann is explicitly utopic: "this study [is] an intellectual or hypothetical one," and "[t]here will be no attempt to state how the required conditions might come into existence, nor to prophesy that they will occur" (Dewey 1927: 333). This is an approach to the problem that is legitimate as well as important but one that lends support to Lippmann's skepticism about the possibility of bridging the gap between experts and citizens.

In my view, the ideal society in question could be reinterpreted—in a way that is not completely unfaithful to the evolutionism-inspired part of Dewey's philosophy—as presupposing a model of what "makes our life worth living" (Nussbaum 1998, 2011; Sen 2000). This ethical standard, in its turn, assumes a communitarian factor whose origin could be identified with a conception of human nature based on shared intellectual and practical problems. According to Dewey, in fact, democracy is not merely a means to reach the end of social stability and agreement but an end as well—"the idea of community life itself," as he puts it (1927: 148). In Dewey's project, it is of essential importance to overcome what he regards as the straitjacket of a democratic society, namely the sufficiency of the majority rule, and therefore of forms of representative democracy characterized by the merely procedural practice of voting. In Dewey's opinion, if citizens cannot participate in public decisions, they are not fully free even when they vote: "Democracy must begin at home, and its home is the neighborly community" (1927: 213), "Unless local communal life can be restored, the public cannot adequately resolve its most urgent problem: to find and identify itself" (ibid.: 216). And then again: "Till the Great Society is converted into a Great Community, the Public will remain in eclipse" (ibid.: 142).

This is a new declination of the ideas of the Swiss philosopher Jean-Jacques Rousseau (1712–78), who, in his *Social Contract* ([1762] 1997), contrasted the sum of the will of the single individuals and their social activities (Dewey's "Great Society") with the General Will (Dewey's "Great Community"). The former can transform into the latter only on the condition that the social consequences of the actions taken by each person are "*known* in the most complete sense of the word" (ibid.). The public is in eclipse until it is subject to the kind of propaganda highlighted by Lippmann; this includes the citizens' ignorance of the social mechanisms of disinformation thanks to which the elite can keep its power. It is only thanks to a new type of scientific education that the public can become aware of these mechanisms.[18]

However, there are at least *three* objections that can be raised against this ideal of democracy. The *first* concerns the fact that, for more than a century now, most of our relationships have become wholly *anonymous*, particularly in big cities, where the fact that we are not known or recognized by a significant number of people has even been regarded as a partial cause for *suicide* (Durkheim 1951). Currently, Westerners do not live any longer in purely rural, isolated, and sedentary societies, and even in poorer countries we do not find the tribes of hunters and gatherers in which "local community life can be restored" but human beings who are dying of hunger and are desperately seeking a better life in richer countries.

Of course, Dewey must have been aware that "face-to-face communication," which is indispensable to form loving relationships and cooperation to reach common aims, can be present only in very small groups of interacting people and not in urban environments, whose members cannot be in constant contact, conversation, or dialogues. Leaving aside the cases of isolations imposed by pandemics, it is superfluous to recall the trite commonplace that today virtual contacts are replacing more and more direct face-to-face interactions. Admittedly, Dewey does not express nostalgia for a time that is lost forever but does not suggest any concrete solution that can help to turn small, cooperative local communities into a very large, citywide, or even nationwide communities. If even very small, more static communities (families!) can be divided by rivalries and hatred that prevent fruitful conversation, how can we construct a very large community of traveling people that interacts in a communitarian and constructive way?

Second, citizens—even if collaborating to solve a common problem and aiming at truth—more than scientists, have *conflicting aims*. From an epistemic viewpoint, the latter aim at truth, certified knowledge and predictive power often as ends in themselves, while the aims and the means chosen by the former are diverse and more pluralistic. Somewhat paradoxically, Dewey claims that the existence of diversified and conflicting viewpoints in the social world gives the public a more important role in deciding policies of common interest. In other words—by encouraging a pluralistic attitude toward practical problems that require conversational practices involving different viewpoints—he argues that the pluralism typical of the social sciences can be of decisive help to bridge the gap between experts and citizens.

Third, Dewey plausibly claims that, collectively, the society or the public at large knows *more* than each expert that is part of it. However, this common knowledge is powerless when the public must take *specific* decisions on highly specialized issues that nevertheless involve their future life. In the social arena, there are pluralistic and diverging interests, different individuals wear different shoes, and the point at which they pinch differs. Given the increasing specialization of science, these cases are becoming more and more frequent. Consequently, the experts must be trusted by both politicians and the wider public as a function of the different roles assigned to them by democratic institutions. This generates the thorny problem of *trust* between the different knots of the social network, one that Lippmann took advantage of by insisting that the public is very often offered distorted images of the social reality.

How could Dewey respond to these three objections? Let us discuss possible replies in turn. As to the first criticism, Dewey's ideal could still be defended, controversially, by *stressing those values that are shared by human beings as such*. Aims like preserving life on our planet, avoiding destructive world wars, achieving a degree of social justice, ensuring a cooperative society, defending human and animal rights, cultivating enthusiasm for truth-seeking enterprises, and artistic beauty can be universally shared—even *without* physical or virtual interaction—with the rest of all human beings. Admittedly, this is in part a weakening of Dewey's communitarian ideals (face-to-face communication) but makes them more realizable.

In order to discuss the *second* objection, it is important to recall that Dewey implicitly tries to answer it by stressing the fact that the interacting mental habits of scientists can and ought to become the reflexive interactive habits of citizens, as if they were all part of a single "community of inquirers" (Barrotta 2017). Dewey's most important legacy is that we should regard democracy as a social enterprise, namely as a way of solving social problems via the cooperation of all actors, and therefore as a collective "form of inquiry" that includes first and foremost scientists and experts, but not just them (see also Bohman 2010: 51). The critical attitudes of scientists entail the acceptance and even the encouragement of dissent and a pluralist attitude and therefore a reasonable dose of skepticism toward the view of the majority.

These epistemic virtues typical of good science are present but ought to be present more in the large public. According to Dewey, cultivating the virtues of critical inquiry (tolerance, openness to doubt, intellectual autonomy) that proved essential for the growth of science must also be relied upon for the improvement of our social institutions. After some time, the vast majority scientific communities reach an agreement over a given theory. Given the inseparable relationship between the epistemology of science and social institutions defended above, the same should happen in a democracy.

One of the reasons why our "approach to moral questions remains mired in traditional, pre-scientific ways of thinking" might consist in the erroneous conviction that the decision to *cooperate* with other individuals is *less* advantageous for the single person than the decision to defect.[19]

The third, "trust problem" can be eased only by increasing the literacy of Dewey's Public. In his review of Lippmann's *Public Opinion*, Dewey writes: "Democracy demands a more thoroughgoing education than the education of officials, administrators, and directors of industry. Because this fundamental general education is at once so necessary and so difficult to achieve, the enterprise of democracy is so challenging" (Dewey 1922: 288). I could not agree more with this quotation, since this is one of the two theses (T_2) that I will defend throughout the book: the experts cannot decide by themselves which pinch is to be given priority.

As a final evaluation of Dewey contributions to the themes I shall discuss, however, I definitely think that he should have stressed more the fact that, on the condition of an increased level of scientific literacy, *participatory and representative* forms of democracies tend to *coincide*. The problem of bridging the gap between experts and citizens denounced by Lippmann can only be alleviated by trying to spread a scientific attitude of mind, which is my thesis T_2. Under this condition, the majority principle, despite the drawbacks that he denounced, turns out to be superior to direct forms of democracy, which is the thesis T_1 that was anticipated in the Introduction.

In sum, at the beginning of his latest book, Kitcher quotes Dewey's *Democracy and Education*: "If we are willing to conceive education as the process of forming fundamental dispositions, intellectual and emotional, toward nature and fellow men, philosophy may even be defined as the general theory of education" (quoted in Kitcher 2021b: ix).

2

How Does Science Work?

The Evaluation and Controllability of Scientific Hypotheses

2.1 Methodological Premises: Descriptive and Normative Aspects of Science and Democracy

Even though it seems overambitious or impossible to try to "define" science and democracy, an attempt to circumscribe, at least approximately, the meaning of these terms will help me to clarify the methodology adopted in this book. The social and cultural phenomena concerning scientific and democratic institutions can be studied from both descriptive and normative perspectives. As is well known, a description uses language to express "how things are" in a selective way and does therefore refer to the sphere of *facts*. The use of language involving norms or values expresses instead what *ought* to be the case, and as such it belongs to the sphere of *values*.[1]

Considering the overall aim of the book, I shall here adopt the latter perspective. Consequently, my main questions will be: how are the values grounding the scientific enterprise linked to those lying at the foundations of representative democracies? Why are achieving scientific knowledge and living in democratic societies both desirable? And finally, why is having or striving for objective knowledge indispensable in contemporary democracies?

The choice of adopting a normative perspective is not an edifying but useless utopian escape in a Platonic world of ideas. On the contrary, it is necessary because, usually, to claim that a hypothesis is *scientific* entails affirming that it is *rational* (in a normative sense) to accept it, so that we *ought* to endorse it by preferring it to competing claims that are regarded as uninformed or pseudoscientific by the scientific community. Analogously, to judge an institution as *democratic* "usually" means to prefer it to other forms of governments, like dictatorships or autocratic regimes, which, for example, systematically kill or torture political opponents and forbid freedom of thought and a free press.

Naturally, we must consider that for some of us not only does science not yield objective knowledge—a thesis that I will show to be false—but also that it is responsible for the irreversible and rapacious exploitation of the planet's resources, with no attention to future generations.[2] Analogue remarks apply to those who are convinced of the superiority of a police state, or of the necessity of paternalistic regimes, where civil freedoms are limited or suppressed.

While granting the existence of these stances—not always minoritarian in Western countries, especially about science—the defense of the objectivity and desirability of scientific knowledge and the preferability of democratic institutions will, in any case, contribute to clarifying the difference between the positive and the negative stances taken towards both science and democracy. Considering that the values grounding the two institutions can be realized to a smaller or greater degree, it becomes very important to find out how we could organize our societies in such a way that a synergy between science and democracy could be established so that a more complete development of human capacities can be achieved.[3]

The first value that I will discuss and that, in its broadest sense, plays a fundamental role both in science and in democracy, is "*controllability*": of hypotheses in science and institutional powers in democracies. Since I begin my discussion with science, in the next section I shall offer a very brief overview of the process of justification of a scientific hypothesis, given that the values of controllability and justifiability in science must be considered together and I will refer to them interchangeably. To analyze the role of the controllability of a hypothesis, I must explain what a scientific inference is and the central role that it plays in the rational justification of our beliefs. I shall begin with very basic information to progressively bring the reader to more complicated arguments.

2.2 Scientific Inferences and the Controllability of their Conclusions

The inferential aspect of scientific knowledge is necessary because the information on the external world that we filter through our senses is restricted to a very limited part of what there is. To give an example, our eyes are blind to the band of infrared (that, however, are perceived by our body as heat), to the ultraviolet ("perceived" by the skin and filtered by the constantly thinning ozone layer), and do not see directly neither the electromagnetic waves captured by our cell phones nor the X-rays used for radiographies. In sum, of all these electromagnetic waves, we see a very restricted band, which, as is well known, is related to our perception of colors (see Figure 2.1).

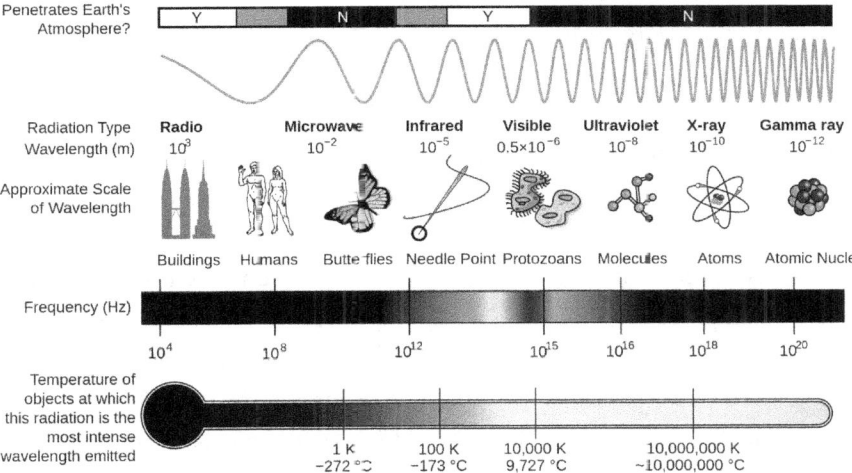

Figure 2.1 The complete spectrum of electromagnetic radiation: on the left of the visible region, there are the infrared rays, and on the right, the ultraviolet ones. The wavelength is the distance between two peaks of the wave, the frequency is the number of complete oscillations of the wave in the time unit. The figure shows that frequency and wavelength are inversely proportional: the higher the former, the smaller the latter, and vice versa. *Source*: Online at: https://upload.wikimedia.org/wikipedia/commons/thumb/c/cf/EM_Spectrum_Properties_edit.svg/2560px-EM_Spectrum_Properties_edit.svg.png (accessed August 9, 2022). Creative Commons Attribution-Share Alike 3.0 Unported license.

The same limited access to the physical world characterizes the frequency of sound: our ears do not perceive infrasound (whose frequencies are lower with respect to the audible ones) and ultrasounds (higher frequencies). The *former*, however, which are originated from brusque air movements, are "perceived" from seismographs, which allow the localization of earthquakes and volcanic eruptions. The latter are "seen" and reflected by the soft tissues of our body.

These simple examples are *prima facie* arguments in favor of the claim that to go beyond the information that is directly grasped by our senses, we need both instruments and scientific hypotheses enabling us to discover worlds that are inaccessible to our unaided sensations. This is possible, on the one hand, thanks to artificial tools that strengthen them (for example, telescopes, and microscopes) and, on the other, thanks to *inferences*. Leaving aside the experimental component of science (which is essential), I shall analyze the concept of inference, because it clearly explains the reliability and the justifiability of scientific hypotheses in general. Furthermore, a discussion of these notions will be important in the next chapters.

Inferences are arguments—either inductive or deductive—that are constituted by premises and conclusions that are advanced with the scope to rationally justify the latter by assuming the former. Deductive inferences mostly characterize sciences such as mathematics and logic, while inductive inferences characterize any science that is based on observations and is therefore empirical.

A well-known example of a deduction is the following: if (a) all human beings have two legs and if (b) Jane is a human being, then the *necessary* conclusion is (c) "Jane has two legs." Therefore, a deduction is an argument that contains some premises—(a) and (b) in the example above—and a conclusion (c), such that *if* the premises are true, *then* also the conclusion is necessarily true. Here, *necessarily* means that it is impossible that the premises of a deductive argument are true and the conclusion false. More generally, if we accept the premises of an argument as true and check the steps of the inference by verifying that it is correct, we must necessarily accept the truth and the validity of the conclusion. A valid deduction transmits the truth from the premises to the conclusion.[4] In deductive sciences like mathematics and logic, the justification of a belief in the conclusion of an argument (generally a theorem) relies on the possibility that anybody has, at least in principle, to *control* the demonstration by repeating it.

Inductive inferences extend and generalize past observations to those that have not already been made. In this sense, they are used to justify predictions, not only scientific predictions but also those on which our life depends. Without predictions, no sufficiently complex animal would survive. Inductions—which characterize empirical sciences like physics, biology, the neurosciences, psychology, sociology, and economics among others—unlike deductions—*do not* preserve the truth of their hypotheses/premises. It is possible that the premises of an inductive argument based on past observations are true but that the conclusion about the future or faraway places is false. A famous example has been proposed by the philosopher and mathematician Bertrand Russell (1872–1970), who explained the essential aspect of inductive inferences. A turkey observes that every morning the farmer gives him food (true observational premises) and it inductively concludes (also animals induce!) that the farmer *will always do so also in the future* (prediction). The proposition in italics is the conclusion of the inductive argument since it extends the knowledge of what the turkey has already observed to what it has not yet observed. The conclusion, however, is false despite the truth of the premises, because the farmer will one day kill the turkey.

In sum, even though many empirical sciences heavily rely on mathematical models, their acquisition of knowledge depends on the formulation of theories that are based on already observed facts that are presumed to be valid, under specific conditions, also in circumstances that have not yet been observed and that possibly will never be. The empirical sciences are fallible, and it is inappropriate for us to say that physics or biology *demonstrate* facts in the same sense in which mathematics does: physical laws may fail to be valid in the future in those domains to which they are thought to apply today.[5]

The best way to characterize the concept of induction is to leave the word to the greatest physicist of the modern age. In the fourth of his *Regulae Philosophandi*, added to the third edition of the *Mathematical Principles of the Natural Philosophy* published in 1726, Isaac Newton (1642–1729) writes: "In experimental philosophy, propositions gathered from phenomena by induction should be considered either exactly or very nearly true notwithstanding any contrary hypotheses, until yet other phenomena make such propositions either more exact or liable to exceptions. This rule should be followed so that arguments based on induction may be nullified by hypotheses" (1998: 796).

Having a predictive character, the empirical sciences must be justified and confirmed by observations or experiments. Observations occur in the final stage of experiments whose results are *publicly controllable* because they are in principle repeatable. This repeatability depends on the reliability of the inductive inferences: lacking refutations of a given theory, if experiments performed in laboratories all over the world and by a high number of different scientists converge to the same conclusion, then the theory is *temporarily* confirmed. "Temporarily" means that, unlike what happens in deductive ones, the empirical sciences are more subject to significant historical changes and, as we will see, even to revolutions that, however, do not threaten their descriptive power.[6]

For instance, after many centuries, Euclidean geometry is still valid, even if only in (infinitesimally) small regions of curved, non-Euclidean spaces. A very small disk circling the North Pole is basically as *flat* as a flat circle in Euclidean space and yet the surface of the Earth is globally *curved*. However, the history of the empirical sciences has taught us that it is always possible that some future experiments or some theoretical revolutions will disconfirm a theory that is now accepted and that will be discovered to be false in certain respects. "Under certain respects" means that scientific theories, even if they underwent radical changes—and even though they contain "errors" that are often eliminated in the development of scientific knowledge—are never dropped like a useless ballast.

In physics, from which all of my examples are taken, there is progress from an earlier to a successive theory when the latter *generalizes* the former, which then becomes a particular case of the latter. This happens when an earlier theory T ends up being applicable only to a more limited domain of phenomena P, while the latter theory T_1 is valid *both* for the phenomena P described by the old theory *and* for new phenomena P_1 with respect to which the replaced theory T fails.

For example, the wave theory of light has generalized the preceding particle theory of light, in the sense that the latter has become a particular case of the former. Light manifests its wavelike character only when the aperture A of a slit through which it passes is *comparable to its wavelength* λ.[7] The particle theory, which represents light as composed of particles travelling in a straight line, is adequate to reproduce the phenomena only when the slit through which they pass is very wide compared to its wave length (Figure 2.2a). However, the narrower the width of the slit, the more evident becomes the diffraction (Figure 2.2b and 2.2c). As is evident from both Figures 2.2b and 2.2c, the corpuscle theory of light is not adequate to describe the observed phenomena: light reaches points that could not be reached if it travelled in a straight line: it is in this innocuous sense that we can say that the particle theory of light contains "errors." The wave theory is valid also for Figure 2.2a, but its effects can be safely ignored. On the contrary the particle theory cannot predict the phenomena represented in Figures 2.2b and 2.2c. In this sense, the wave theory is more general than the particle theory, since it is experimentally adequate in all three cases.

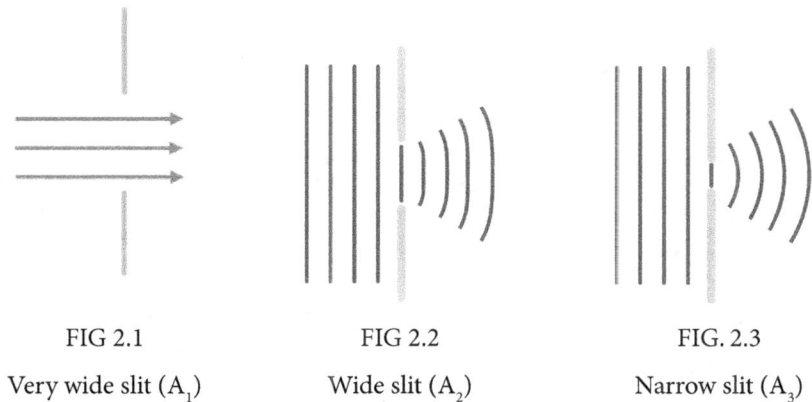

FIG 2.1   FIG 2.2   FIG. 2.3

Very wide slit (A_1)   Wide slit (A_2)   Narrow slit (A_3)

Figure 2.2 The phenomenon of diffraction, describing the passage of a light wave through two slits of different aperture, A_2 (center) and A_3 (right), with A_2 wider than A_3. If light were composed of particles propagating in a straight line (as in Figure 2.2a), it would be impossible to explain why it also reaches points that are off the trajectory that is orthogonal to the upper and lower point of the slits. The diffraction becomes more pronounced when the slit narrows (Figure 2.2c on the right), while if the slit were very large, the phenomenon of diffraction would not be noticeable but would still be present and light could be treated *approximately* as composed of corpuscles travelling in a straight line. The wave theory is valid in all three cases. *Source*: Figure 2.2a © Mauro Dorato; Figures 2.2b and 2.2c © Bernard Burchell. My thanks to Bernard Burchell for his authorization to use Figures 2.2b–c.

Analogously, on the one hand, it is possible to claim that Newton's theory of gravitation—which postulates the existence of an attractive *force* between any two massive bodies—is false. After Einstein, we know that this force *does not exist*. To explain the fact that smaller masses are "attracted" by larger ones—the Earth "is attracted" by the Sun, or the Moon by the Earth, etc.—the general theory of relativity assumes that any massive body curves spacetime in the same way in which an elastic carpet (representing spacetime) is bent by a stone. The greater the mass, the greater the deformation in spacetime (the carpet).

At the center of Figure 2.3, the Earth bends spacetime with its mass. The Moon (not represented in the picture) rotates around the curvature/valley generated by Earth's gravitation field and continuously rotates around it due to the greater deformation caused by our planet on the carpet. On the one hand, the Moon spirals around the Earth because it tends to fall toward it. On the other, it does not fall onto the Earth because it moves in empty space and keeps its inertial velocity v along every point of its trajectory. The change in the direction of the velocity represented by v generates an acceleration of the Moon toward the Earth.

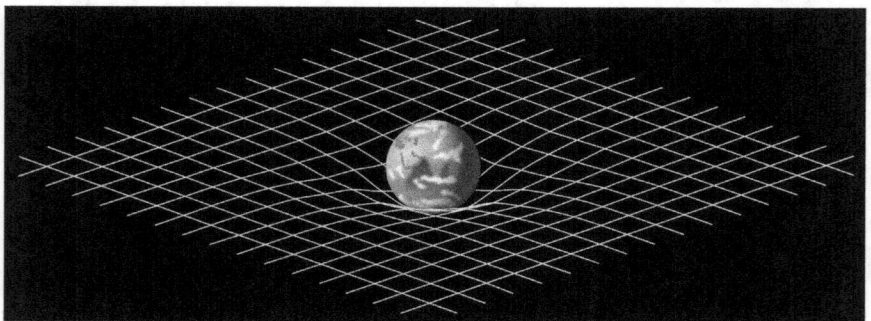

Figure 2.3 The spatiotemporal curvature explains the motion of the gravitating objects: the Earth represented in the picture determines a deformation of the elastic carpet (spacetime) because its mass is greater than that of the Moon (not represented in the picture). Both masses generate a curvature in the carpet, but the deformation produced by the smaller mass of the Moon is negligible with respect to that produced by the greater one produced by the Earth. Consequently, the smaller mass will circle the greater mass, *as if* it were attracted by the greater mass. *Source*: "Gravity," Wikipedia, online at: https://en.wikipedia.org/wiki/Gravity#/media/File:Spacetime_lattice_analogy.svg CC BY-SA 3.0 (accessed August 9, 2022).

On the other hand, however, Newton's theory of gravitation is not "entirely" false, but *false only under certain respects*, since in the sense above it is a particular case of Einstein's theory of gravitation and is valid and applicable when the gravitational field is very weak. Newtonian mechanics is still useful to send a satellite into orbit. Einstein's theory, on the contrary, holds *both* when the gravitational field is weak and when it is strong.

2.3 The Fallible Character of Science and its Social Consequences

The two examples show in a clear way that the discovery that a scientific theory can be found to be wrong in certain respects will seldom force us to eliminate the refuted hypothesis. Moreover, this brings about a great advantage that is both theoretical (in science) and practical (in everyday life). The *theoretical* advantage consists in the drive toward the discovery of a new theory that allows us to eliminate "errors" while retaining the consolidated preceding hypothesis: human knowledge has developed and still develops thanks to the elimination of partially false theories, which are substituted by observationally more accurate experiments. The *practical* advantage is linked to the fact that our beliefs are a guide to action, and having false beliefs is the surest way to fail to reach what we want: if my belief that there is water in the fridge is false and I have the desire to drink, I will not quench my thirst.

Since a quick elimination of a false belief is advantageous, the epistemic attitude that characterizes a scientific community is the openness to criticism and doubts. Naturally, the members of a scientific community typically do not believe or doubt *all* hypotheses, since both epistemic attitudes would not be constructive. A good scientist is especially capable of moving her research toward obscure areas of contemporary research namely those whose foundations are still unknown or not solid. For this reason, the great mathematician and philosopher Henri Poincaré (1854–1912) claimed that: "to doubt everything or to believe everything are two equally convenient solutions; both dispense with the necessity of reflection" ([1902] 2017: xxii). Furthermore, both attitudes have disastrous practical consequences: the former is an over-cautious but paralyzing defense from the adoption of many or all false hypothesis, the latter a dangerous lack of skepticism whose only advantage is the adoption of many or all true beliefs.

It is for this reason that the scientific communities, even though they take for granted many scientific beliefs, are in any case aware of the fact that consensus can be reached only through an education to criticism and acceptance of different points of view. It is the opposite of what happens in the so-called *closed* societies and communities, which are based on a dogmatic knowledge and *static* religions and morals.[8] The view that *science* is not synonymous with *certainty* is now a widely shared, and a more general fallibilism stance about our knowledge, generally attributed to the philosopher of science Karl Popper (1902–94), had already been clearly stressed by the great logician, mathematician, and philosopher Charles S. Peirce (1839–1914), the father of American pragmatism.

Of course, a philosophical thesis cannot be considered true only because it has been defended by a great philosopher. Given its argumentative strength, however, it is very tempting to report the following long quotation from Peirce:

> infallibility in scientific matters seems to me irresistibly comical, [...] In those sciences of measurement which are the least subject to error – metrology, geodesy, and metrical astronomy – no man of self-respect ever now states his results, without affixing to it its *probable error*; and if this practice is not followed in other sciences, it is because in those the probable errors are too vast great to be estimated [quoted in Buchler 2014: 3]. [...] I used to collect my ideas under the designation fallibilism; and indeed the first step toward finding out is to acknowledge you do not satisfactorily know already; so that no blight can so surely arrest all intellectual growth as the blight of cocksureness [ibid.:4]. [...] Indeed, out of a contrite fallibilism, combined with a high faith in the reality of knowledge, and an intense desire to find things out, all my philosophy has always seemed to me to grow [ibid.].

John Stuart Mill (1806–73), a philosopher who was also very attentive to the interaction between science and democracy, analogously wrote that the only epistemic warrant for a scientific hypothesis is that it has not yet been refuted.

This mental attitude of openness to doubts is an important inheritance of ancient Greek philosophy, expressed by Plato in particular in his Socrates' apology, where he has Socrates claim that he, Socrates, is conscious of his complete ignorance and for this very reason, according to the Delphi oracle, is the wisest of all men. This form of *docta ignorantia* (i.e., *learned ignorance*)[9] has characterized the most creative moments in the history of philosophy and science. Note that Peirce gives an important twist to the Socratic words: rather than saying that "we know that we do not know," he says we know that "we do not know in a satisfactory way."

In any case, the fallibility of scientific knowledge has very important consequences with respect to the values that are at the basis of democracy. The tolerance of other people's opinions (even when we consider them to be mistaken) is based on the awareness that our beliefs could be mistaken, too. In a certain sense, tolerance is grounded on our ignorance and the possibility of errors: "What is then tolerance? [...] It is the consequence of humanity. We are all formed by frailty and error, let us pardon reciprocally each other's folly,"[10] as Voltaire (1694–1778) wrote in his entry *"Tolerance"* in the *Encyclopédie*. Fanaticism is instead the contrary of tolerance since it is based on the thesis that our beliefs are infallible: to quote again the master of tolerance: "Fewer dogmas, fewer disputes; and fewer disputes, fewer miseries" (Voltaire [1763] 2000: chapter 21, p. 42).

2.4 The Controllability of Scientific Knowledge as a Social Phenomenon

To prepare the ground for highlighting additional close analogies between the procedures grounding both science and democracy, it is important to stress the following point. Since the progress of knowledge is based on the elimination of errors, the creation and above all the control of a scientific hypothesis are essentially the product of a very complex social and selective process. In a very precise sense, it is the *whole* community of experts, or at least a great part of it, which judges the acceptability of a new hypothesis, and it is only at the end of a long social process of critical evaluation that a hypothesis can be considered as "accepted."[11]

Therefore, scientific fallibilism must be understood in the right way. If an individual scientist (or a small group of researchers) working on the solution of a problem, tried to prematurely falsify findings that are still partial—as Popper's naive fallibilism at times seems to suggest—no well-confirmed hypothesis will ever be arrived at. It is only after the publication of a new hypothesis that the criticism of the peers typical of the scientific method can intervene, and it is only at this stage that the theory can be amended thanks to this criticism.

To understand the influence that scientific communities have in the formation of consensus, it is sufficient to consider the following fact. The elimination of errors during the attempts to solve a scientific problem and the discovery of scientific frauds can happen only thanks to experts, who have worked in the relevant field for a long time, and not thanks to citizens who did not spend a great part of their life to acquire the necessary knowledge. Moreover, we must consider that, on the one hand, in scientific institutions there is an incentive to the discovery of errors: this in fact accrues to the scientist's reputation. On the other hand, since it is not possible to presume that all scientists have a moral code that is better than that of the rest of the human beings, scientific institutions have created mechanisms discouraging cheating with the worst social punishments, that for a scientist is the loss of her peers' respect.

To have a rough idea of how the social control of a scientific hypothesis works—and then inquire whether a similar procedure is applicable also in democratic institutions—we should imagine a group of scientists who wants to publish the results of their work in specialized journals. The best journals typically use peer reviewers selected from the editorial board for their recognized expertise and reputation. In many journals, the paper is evaluated in such a way that the author(s) ignore(s) the identity of the referees and conversely, the latter, except for books, ignore(s) the identity of the former. This "double-blind" system is sometimes strengthened by a third rule: the editorial board ignores the identity of the author who submitted the paper until the end of the process, in such a way that the decision to publish the paper is not influenced by the prestige of the academic institution to which the author belongs.

Despite these procedures, sometimes the reviewer identifies the author's identity. Suppose that a reviewer receives a paper that has been written by a colleague whom she knows well and that she thinks that this fact can influence her verdict: in this case, she has the professional duty to inform the editorial board. Sometimes, this rule is violated but this does not entail that the evaluation of papers that are submitted to journals with a very good reputation is not impartial in most cases. This impartiality has been criticized: especially in certain disciplines, is it not the case that the *same facts* can be interpreted and explained in different ways? However, "respecting the independence of facts" is the first step toward any possible scientific discovery. If there were no difference between a fact and its interpretations, what would the interpretations be *about*? And without independent facts, would not these interpretations be completely unjustified? As we shall see in more detail Chapter 7, any interpretation of a fact always presupposes that there is something to be interpreted that is different from the interpretation. Once a scientific hypothesis is confirmed, it is not possible to "quarrel with facts." "We are entitled to our opinions but not to our facts."[12] Consequently, not only does science presuppose that there is a "way in which things are" that does not depend on anyone's will, but it also reinforces the corresponding belief in the society at large, a belief that is not incompatible with fallibilism as described above. The progress of scientific knowledge can be described as an increase in the capacity to shed more light on "what there is."

It is essential to note that the "dialectical" exchange typical of the scientific method occurs *before* submitting a paper for publication: colleagues and other peers read the paper and offer critical suggestions thanks to which the author corrects her theses or eliminates possible errors. However, it is in the reviewing stage taking place *after* the submission that the authors' main claims are severely criticized and evaluated. The reviewer must try to find out weak points in the authors' theoretical hypotheses, in the experimental results, or in the proof of a theorem. This selective process ends up with a "reject" or an "accept" or "accept with *minor* or *major* revisions." We should keep in mind that the best journals publish only 5 to 10 percent of the submitted papers,[13] and that those that are published have been strengthened and made more convincing by the critical remarks that anonymous referees have sent to the authors. It is very rare that the published version of a paper coincides exactly with the draft that had been originally submitted. It is not too rare that after the publication, other authors reply to the original paper by writing another one that criticizes it in the same or other journals. A reply to this criticism, if justified, can be published, and the process stops when a vast consensus is reached.[14]

Furthermore, consider that the publication of a new hypothesis in a journal or a book is not the only stage at which the members of a scientific community evaluate a new theory. When a new theory is presented at a conference, for instance, the peers typically criticize the new thesis, which often "forces" the speaker to defend it as well as possible by answering the objections raised during a *public discussion* that takes place often in front of a large audience. If the presented thesis is not robust enough to resist the audience's objections, the speaker must either correct it or later abandon it.

It is worth noting that also philosophy, given its vocation to engage in critical analysis, has played and still plays an important role in probing the conceptual foundations of scientific theories. Our most fundamental scientific theory, namely quantum mechanics, is a case in question. In 1927, the fifth of the "Solvay" conferences had as speakers, both Niels Bohr (1885–1962) and Einstein, who publicly debated on the conceptual implications of the new theory for our knowledge of the physical world. These discussions involved also other founding fathers of the new theory (see Bacciagaluppi and Valentini 2009) and were thoroughly *philosophical* in nature, even if the protagonists, strictly speaking, were physicists. Such debates are still very influential in contemporary philosophy of physics and are of great importance for at least two reasons. First, they promoted and promote to our days a deeper understanding of the *meaning* of mathematical formulas that, even if unparallelly accurate and effective in predicting the relevant phenomena, cannot still answer exactly the key question: "What is quantum theory about?" Second, the essentially philosophical and conceptual questions that Einstein in particular raised at that conference have remained lively also after 1927, since they have played and still play an important role in contemporary technological developments, from quantum cryptography to quantum computers.[15] Widely discussed philosophical questions can be of heuristic value not only for theoretical science but can also stimulate technological developments.

Another important social aspect of science is centered on scientific academies, which in the sixteenth and seventeenth centuries accompanied the formation of the modern European states and had an important role in the diffusion and the internationalization of science.[16] Thanks to the publication of the Proceedings of the Societies' meetings, scholars who belonged to different national Academies communicated the results of their research also to foreign members of other Academies. Proceedings are not just the progenitors of contemporary scientific journals but are still a publishing form of scientific workshops (the so-called special issues). These types of publication made available to a larger number of scientists the results found in one Academy, thereby making possible an independent, additional check of the relevant hypotheses.

The societal and political consequences of these exchanges can hardly be exaggerated. Given that the publication of scientific results was addressed also to scientists coming from different countries, since its origin science had a very important "international" role, made possible by its *cosmopolitan* nature,[17] based on observations and common methodological assumptions. Science textbooks used by students in London, Beijing, and Singapore, except for their language, have the same conceptual content.

Thanks to the objective character of scientific knowledge, the relations among scientists coming from different parts of the world have contributed and can contribute to spreading a supranational ideal that helps to avoid dangerous forms of populistic nationalisms. By becoming more literate in science, human beings presumably would feel to belong to the largest human community and therefore progressively develop the conviction to be "citizens of the world." In addition, they could also be more prepared to fight against those ideological distortions of science that in the last century took place in Germany and the Soviet Union. For instance, the recent discovery that the concept of "human race" has no scientific basis is just one example of the beneficial aspect of science.[18]

Finally, to highlight even more the social character of science, I cannot omit mentioning epistolary and letters that have been exchanged by scientists, which, at the time of the internet, a historian of science can no longer use. Two of the most important historical examples of a discussion of a scientific hypothesis are the letters on the nature of space and time written by Gottfried Wilhelm Leibniz (1546–1716) and the theologian Samuel Clarke (1675–1729) who defended Newton's natural philosophy (Clarke 1717), and the letters that two of the founding fathers of quantum mechanics, namely Max Born and Albert Einstein, wrote on the interpretation of this fundamental physical theory (see their letters of 1916–55 in Einstein, Born, and Born 2004).

In short, the formation of scientific consensus, which originates thanks to a discussion among experts, should ideally influence also a community of well-informed citizens that initially disagrees on how to solve social problems.[19] The values of openness to critical discussions and respect for facts—which are typical of well-organized science—have been and can be in the future an important source of inspiration for all citizens living in a democracy, provided that their level of education is sufficient to realize both the main aspects of the problem and the economic interests that can lie behind certain hypotheses.

2.5 The Diachronic Aspect of the Control of Hypotheses and the Evolution of Social Norms

In the previous section, we have seen in what sense the procedure of constant control of scientific hypothesis, favored by an attitude of "organized skepticism,"[20] is a necessary condition for the advancement of knowledge. This suggests that such a system of intellectual norms could be studied not only from a synchronic point of view but also from a diachronic and intergenerational perspective.

In this respect, scholars such as Popper (1972) and Hull (1988) have suggested some important analogies between the evolution of biological species and the evolution of scientific knowledge. In short, the former kind of evolution is based on two mechanisms: (1) a blind *variation* of the transmission of the genetic heritage among individuals belonging to the same biological species; and (2) the natural selection of individuals and species that end up not being able to reproduce. Cultural evolution as presented above has striking similarities with these two mechanisms. On the one hand, there is, (1) the cultural transmission of scientific beliefs from one generation to the next (via a continuous process of *learning*) always presents some *variations* with respect to what has been learned. On the other hand, there is, (2) the selection of such beliefs with the elimination of "errors" (biological species in the analogy), that is, of hypotheses that have been falsified by data and that are no longer taught by the relevant social institutions.

The obvious difference between the biological and the cultural evolution depends on the nature of the mechanisms that are responsible for the transmission of variable information: a genetic and therefore blind *chemical-physical* apparatus, on the one hand; a method that is *cultural* and oriented toward *consciously pursued aims* to reach the truth or the empirical adequacy of the hypothesis, on the other.[21] However, there are striking similarities: the role of natural selection, which prevents some species from reproducing in modified environments, is implemented in science by experiments and observations, which let theories that are in accord with the hypotheses survive but eliminate those that are inconsistent with them. As Popper wrote, in science, "we let theories die in our stead." Moreover, as a third analogy, we could consider the fact that exactly as the reproductive success of a species' genes favors its survival, the reproductive success of a cultural unit of transmission, that Dawkins (1989) called *meme*, "infects" more brains by making its survival across different generations more probable.

In philosophy, the analogies, even when they are as suggestive as this one, are dangerous and must be handled with care: after rigorous analysis, they often reveal only a superficial resemblance. However, *two* considerations suggest that, in our case, the analogy between biological and cultural evolution is instructive as well as strong. The *first* is that biological evolution suggests a winning strategy also in scientific changes: the variability of the genetic heritage from one generation to the other helps the survival of the species. In modified environments (say, when birches around which white butterflies fly are blackened by pollution), black butterflies belonging to the same species can camouflage better, since the white ones become more visible to the predators. Analogously, the pluralities of scientific hypotheses, especially in controversial areas of scientific research, can help the solution of open problems by increasing the chances of "survival" of at least one theory, the one that eventually will pass the experimental texts.

The *second* consideration is suggested by the fact that natural selection has occurred also between *groups* and not only among individuals (Sober and Wilson 1998). Groups in which individuals of the same species have cooperated reproduced more than other groups in which individuals maximize their profit by exploiting the cooperation of others (the so-called *free riders*, those who do not pay the bus ticket). Human beings, unlike bears, are social animals. There are studies, which cannot be discussed here,[22] that use a Darwinian selective model to explain in a formally rigorous manner the origin of the social contract. Such a contract is an idealized and normative model of great generality which does not faithfully reflect historical processes but is advanced to explain why it is better for originally "free," isolated individuals to accept a set of rules for living together, which are binding and advantageous for everyone. In fact, emerging spontaneously from selective processes, this cooperative model is more efficient for the achievement of the goals of the individuals than any other kinds of interaction among them.

We can conclude by saying that the cooperation typical of scientific communities that reached a consensus after an original disagreement ideally may be regarded as corresponding to the cooperation among individuals in democratic societies that are initially in a state of competition and conflict about the solution of some social problem. Each of these individuals recognizes the necessity to respect some key rules that are binding for everyone, and that are more efficient to promote individual goals that may originally diverge. The fundamental point is that, in an abstract but formally precise way, it is possible to formulate models that determine the optimal equilibrium between conflict and cooperation in a group of animals, and between consensus and disagreement in a community of both scientists and citizens belonging to a democratic state. A dialectic between ideas, despite the different forms that it takes in the two communities, is a form of cooperation, and has as its final scope the achievement of a more stable consensus. However, what does "in different forms" exactly mean?

2.6 The Role of Criticism in the Decisional Processes of a Democracy

To answer this crucial question, it is appropriate to investigate the role that the controllability and the criticism of problem-solving hypotheses—the fundamental elements making scientific progress possible—play in the political decisions of a democratic community. In this respect, the following *four* considerations seem rather plausible:

1. We have seen that the acceptance of a scientific hypothesis presupposes a socially complex procedure and a collective decision based on a critical

examination carried out by experts. Also, the decisional processes of a well-functioning democracy are based (but should be based even more) on public discussions, which are open to criticism coming from the opposition, a free press, books, associations, parties, labors, etc. What happens in open discussions during a scientific conference corresponds and ought to correspond to the role of public opinion in democracies, where groups of citizens with different ideals and interests can freely express their opinions.

Note the following close analogy. On the one hand, scientists check and criticize the results reached by their colleagues to increase their (and our) knowledge by formulating new theories. On the other hand, the "checks-and-balances" system that characterize an ideal democracy—the well-known separation among the three powers (judiciary, legislative, and executive) advocated by Montesquieu in 1748[23]—have the function to criticize decisions that, qua expressions of conflictual interests, do not necessarily favor the common good. In any case, the pluralism of interests that must characterize a liberal democracy presupposes a shared, unanimous acceptance of overarching rules of the games ("the universal rules" of democracy), first and foremost the possibility to replace a freely elected government with another. Such a replacement presupposes in turn a pluralism of parties and is justified when the elected party fails to solve the problems for which it had been voted. Of course, the consensus characterizing scientific communities cannot be compared to the pluralistic conflict of interests featuring in a democratic institution. However, in those cases in which important problems of common interest have not been adequately solved, the democratic procedures allowing to substitute one party with another are not too dissimilar from those characterizing scientific change. In both cases, some rules must be respected that maximize the chances of success of a problem-solving hypothesis.

2. The controllability and repeatability of the observations allowing confirmation or rejection of scientific hypotheses have crucial political consequences. Access to knowledge and information is one of the main sources of power. To the extent that *every citizen* in principle[24] has the possibility of controlling and criticizing—albeit in a necessarily indirect way—the validity of a scientific hypothesis, the ideal that all human beings possess equal rights is strengthened. Conversely, the principle of equality of rights on which a

democratic society is based favors universal access to knowledge, which is a necessary condition for its growth.

3. On the contrary, in those societies in which the access to knowledge is limited by someone who allegedly has special or closer contact with allegedly indubitable sources, there will be a very strong tendency to an unequal distribution of rights and political power. In these circumstances, non-democratic societies will prevail, where neither the deliberative principle based on the majority rule, nor the system of "checks and balances" between different political powers can be practiced. Note that this really happened in the past: in many societies or tribes based on sacredly mysterious, cryptic, and in any case non-controllable knowledge, or on magical visions of the world, dominant casts justified their power by making people believe that they had a special relationship with one or many supernatural powers.[25] Moreover, the constructive, skeptical attitude that science has cultivated since its beginning has been the best antidote against the conformism that dictatorial regimes tried to inoculate with propaganda and terror. For example, during the Soviet dictatorship, courageous scientists fought against a tyrannical power, as the telling example of the great physicist and laureate of the Nobel Prize for Peace, Andrei Sakharov (1921–89), shows.[26] The pluralism of scientific hypotheses that characterize different research programs in periods and areas where universal consensus has not yet been reached is fundamental also to generate progress in social problem-solving when there exists disagreement about the aims and the means to reach them. A democracy grounded in a scientific approach to knowledge should (a) encourage and defend the emergence of new proposals for solving urgent social problems advanced by individual citizens or, more realistically, by groups of citizens, and (b) guarantee that such proposals can be critically discussed by other citizens without jeopardizing more or less universally shared rules of the game.

4. We have seen why the critical and collective controllability of hypotheses are the fundamental characteristics of all sciences and why the deductive control and the empirical tests give us strong reasons to believe any adequately formulated scientific hypothesis. Analogously, justifying to the citizens some reforms aimed at solving social problems is essential for the well-functioning of democratic institutions. *Justifiability is a fundamental ethical value in all human practices, for it is linked to the responsibility of the opinions that we*

adopt. Etymologically, responsibility comes from the Latin *respondēre*, which entails "to be ready to answer questions like the following": "For what reason should I believe in the same hypothesis that you believe in?" "Why do you believe that this new law would favor the common interest?"

To sum up, I have highlighted how the ideal of the universal *controllability* of scientific knowledge could strengthen and be strengthened, as we will see in the next chapter, by the equalitarian ideal that is characteristic of democratic societies. Due to the exponential growth of specialized knowledge, such controllability is realizable only *in principle*. For instance, only those citizens who dedicated a great amount of their time to the study of the human immunological system are capable of empirically controlling the hypotheses they defend and therefore of expressing an authoritative opinion to which we should defer.

In other words, the exponential growth of the specialized knowledge that we can see under our eyes makes the control of individual hypotheses *unrealizable* by individual citizens. More realistically, we must accept that the control of the validity of scientific hypotheses must be *mediated* by experts, a fact that implies that, on our part, such control must necessarily be *indirect*. Anticipating some issues to be discussed in more detail in the Chapter 3, it is the indubitable growth of specialized knowledge that constitutes a very strong argument in favor of my thesis T_1: that is, in favor of the preferability of *indirect* and *representative* forms of democracy—where electors vote for their representatives—to forms of democracy where citizens are very often *directly* involved in decisions concerning many issues of social interest.

This fact is crucially important for my discourse, for it automatically raises an issue that can be anticipated also at this point. The imperative "delegate to those who know," which is requested by the increasing specialization of knowledge, necessarily entails a degree of *trust* in the experts that, however, should not lead to technocratic forms of government. In a technocracy, citizens would have no decisional autonomy, for they could not contribute to establishing the aims of the instrumental use of the technical knowledge that only the expert has. The problem is made more serious by the fact that the elected politicians who, as should be the case, are more competent in social policies than most citizens, have often to rely in turn on experts whose technical knowledge is far more specialized than theirs. Before discussing this difficulty, however, I must clarify in the most schematic, but as precisely as possible way how "democracy works" or, as modestly as possible, what links the principle of equality, the principle of majority, and the principle of the separation of powers to the values that are at the basis of the scientific enterprise that I discussed in this chapter.

3

How Does Democracy Work?

The Balance of Powers

With inevitable simplifications, in this chapter I will briefly deal with the universally known, essential principles, that characterize democracies: the principle of equality, the principle of the majority (understood as an essential rule for the actual implementation of the people's government), and the principle of the separation of powers. The novelty of my discussion lies in the attempt to yield a deeper understanding of the relation linking these three principles with the social norms mentioned in the previous chapter and that ground the epistemological endeavor of science.

3.1 Democracy and the Equality Principle as Founded in the Free Access to Scientific Knowledge

From our daily experience, we know too well that making decisions is an inevitable and often painful feature of our life. Quite literally, we are always forced to decide: even the decision of not making any decisions (by delaying them) is itself a decision. Analogously, the first problem that a community of people who live in a democracy must face is how to decide between people who are "equal" in the sense to be explained below but do not necessarily share the same ideals and interests.

In the Jewish-Christian tradition, the importance of equality among human beings has a religious derivation and justification: in the American Declaration of Independence (1776), for instance, we read that "all men were *created* equal." However, since this value is an essential element of all democratic constitutions, nowadays it must be justified independently of any explanation that goes back to its religious origins. To clarify this problem further, we had better start with this question: Under which aspect are we "all equal"? And according to what?

Even if all the members of our species have the same anatomy and very similar cognitive faculties, in democratic institutions referring to the value of equality implies that we put aside, as irrelevant, facts relative to any difference related to height, physical strength, the color of our skin, or intellectual capabilities (even assuming that the latter could be measurable, as memory possibly is). We are equal only before the law—"without any distinction of gender, race, language, religion, political opinions, personal and social conditions."[1] This is because the concept of "equality before the law" depends on, and is warranted by, the fact that every human person enjoys the same "unalienable rights." In the Declaration of American Independence, this concept is normative because such are the human rights that ground it. It is for this reason that one of the fundamental tasks of a democracy is to make sure that these rights are safeguarded since they are unalienable and uninfringeable. In the liberal tradition typical of the "state of law" (also known as the "rule of law"), the actual existence of human rights does not depend on the existence or our belonging to a state: on the contrary, the state exists to protect them.[2]

Any reference to the normative aspect of the notion of equality explains the fact that democracy is somehow an ideal to which we must aspire; this also justifies the normative approach declared at the beginning of the first chapter. As with other ideals, also this one can be at least partially realized. This is exemplified in the American history of the extension of the voting rights from white male citizens of the upper social class to white male citizens of a lower class, and then from white women to black men and women which is the history, as far as the voting right is concerned, of the progressive extension of the ideals of equality (see Kitcher 2021a).

For my purposes, it is here important to notice that the notion of equality of rights has developed at least partially in relation to the epistemological model discussed in the previous chapter and which has become dominant since the birth of modern science. The rationality of the scientific attitude, in fact, is the actual expression of our shared rationality and makes possible both the universal accessibility to the sources of knowledge and the control of the validity of scientific hypotheses. Consequently, despite the inevitable differences in talent and predisposition in each of us, from a legal viewpoint all human beings are equipped with the same capacities and the same rationality.

It is also interesting to notice that, together with the formal equality (equality before the law as well as enjoyment of the same rights), after the quote above the third section of the Italian Constitution refers to a more *substantial equality*, based on which the state "has the task of removing the obstacles, of financial and social nature, which, by limiting the freedom and equality of the citizens, prevents the full development of the human person and the actual participation of all the workers to the political, financial and social organization of the country." Even this second kind of equality has important consequences on the ethics and politics of science and therefore deserves some additional comments.

The deeper meaning of the second part of the third section of the Italian Constitution is that the inequality in the starting conditions between individuals, both social and financial, makes it much more difficult that the potential of those who are more disadvantaged can be actualized. Not only is such an initial disparity unfair from an ethical point of view and for the involved individuals, but it is also unfavorable for the collectivity: the full development of the capacities of citizens who are disadvantaged at birth would have contributed to the overall well-being of society. That the *substantial* equality, as the formal one, is an ideal to which we must aspire became explicit with the French Revolution and was proposed against the idea that the social prestige of an individual depended exclusively on birth and not on merit.

The legal bond and the ethical obligation to reduce the disparities in wealth make possible the fact that the merit, the effort put down in work as well as the achieved competence differentiate over time and in a more justifiable way individuals who, at birth, should have enjoyed comparable starting conditions. What the nineteenth-century liberal socialism labeled "the principle of starting gate" or "the fair run," cannot be realized unless the starting blocks are sufficiently aligned. Is there anything in the social norms of science that corresponds to this way of declining equality?

The reason for which, from a perspective comparing scientific and democratic values, even the substantial version of the equality principle is meaningful, lies in the fact that we should favor a fair competition between two or more rival research groups that need public funding and equal access to research data. With no aim to glorify the ideal of the "free market of ideas," the competition between scientific theories is important because, with the help of different and pluralistic approaches, it can stimulate the creation of new avenues of research by disseminating its seeds in different directions Feyerabend (1975).

However, as in the case of the substantial equality between individuals, the competition between research programs can be fair only if the funding process does not *a priori* eliminate the possibility of funding another, competing project, even if the latter could well be the one, among the two projects, that temporarily lacks any immediate technological application. Also in this case, the criterion of social utility requires that both research programs should be given a chance, so that they both get funded (even if, possibly, in a different way).

I can thus highlight the following analogy. For the society at large, the substantial and not merely formal equality in the starting conditions is justified by the fact that, had not the citizen started further back, not only would her talent have developed more fully (even in comparison with those who were more advantaged at the beginning), but also—and exactly for this reason—she could have made it available to the wider society. Analogously, the history of science has shown that we cannot exclude that research project A (at the beginning less "technologically promising" and therefore "more theoretical" than project B, and for this reason possibly rejected for the lack of funding only granted to B), if appropriately financed, would have contributed as much as B or even more to the well-being of our society, to the growth of knowledge, and to technological spinoffs.[3]

It must be noted that there is a significant difference between the two kinds of distributive justice. In the case of the individual citizens and the society to which they belong, the ideals of justice and social equality are much more important than those that aim to fund the possibly only temporarily less applicable scientific projects. In both cases, however, distributive justice favors farsighted and fair decisions that our society ought to make on its future. However, in both cases what is implicated is the necessity that access to scientific and social information be open. In the case of science, in fact, two competing research groups can contend equally only if, from the very beginning, the empirical findings are made available to both. Unfortunately, to keep the results of the research secret or reserved is a consequence of the subjugation of science to the interests of the market, of groups, both military and industrial, or of important pharmaceutical firms. All this contrasts with the cosmopolitan and supranational ideal that characterized the birth of modern science sketched above, that is, with the importance of announcing publicly all the results that scientists have gathered as well as all the scientific hypotheses that they rely on.

3.2 Democracy as the Government of the People: A Comparison with the Majority Principle in Science

It would not be worth recalling the second feature of a democratic institution because it is known to everybody, namely that a democracy, even etymologically,[4] is the government of the people, if it were not so important both for questions related to populism and for its relationship with the way decisions are taken in scientific communities. "People" is a term that needs some disambiguation: on the one hand, the people carry their sovereignty in the ways dictated by the Constitution. On the other, however, given that, on the basis of some indisputable experience, the people are not formed by *a shapeless mass of individuals with the same interests and ideals*—we have seen that human beings are formally equal only from the point of view of the rights that they have before the law and of their common rationality—the ideal of the government of people justifies and makes it necessary that the decisions of a community are taken by following the majority rule.

In principle, there is no reason why the opinion of the majority is both ethically and theoretically preferable to that of the minority. In 1933, Hitler took power thanks to the free vote of most Germans. However, the rule that decisions should be taken by following the majority has some plausible motivations, provided that citizens have a high level of education, as well as, relatedly, the relevant information to make rational choices. However, this is not always sufficient, given that, for instance, a significant number of ranking Nazis had higher education, all the way to PhDs.

In a very important sense, what we call "objectivity" is based on the highest degree of rational agreement between different individuals, each equipped with sufficiently relevant and reliable information. In our case, the agreement concerns the solution of those social problems that are regarded as more pressing by most citizens.

Moreover, the value of social cohesion, which is potentially favored by the rule of the majority, has also a relevant evolutionary meaning given that, since the remote past, the survival of the human groups required that the greatest number of people cooperated with each other and identified themselves with the preferences of the majority. It follows that, to favor social cohesion, the decision taken by the majority had to be binding for all the members of the community and, hence, also for the dissenting minorities.

From the decisional role that the principle of majority plays in science, however, a conflict emerges with the view so far defended about the dependence of the values at the root of democracy on the values grounding science. There is in fact a clear difference in the ways in which in the two institutions consensus is reached. Although it is true that in science the principle of majority seems to be playing a more limited role, this claim needs to be more carefully examined.

On the one hand, one must recognize that the decision to regard a scientific hypothesis H as corroborated by the experiments has nothing to do with how many scientists believe H. Just to give an historical example, not only were Copernicus (1473–1543) and Galileo (1564–1642) obviously right against the opinion of the great majority of non-scientists, but also against the opinion of experts who worked in the same field (astronomy). In other words, the fact that a hypothesis H is supported or not by a set of empirical data E is a logical question that has nothing to do with the number of people who believe H. In such cases, the logical question has the binding feature of an epistemological norm that also has practical consequences. The meaning of "logical" here can be illustrated by the following example. If the probability of rain in the next hours (hypothesis H), given the data on air pressure, temperature, winds, etc. (evidence E) is between 80 and 95 percent, based on E it is both rational and correct to accept hypothesis H and bring with us an umbrella, independently of the number of people who share the hypothesis by knowing about it.

On the other hand, however as we have seen in the preceding chapter, the decision to accept a scientific theory is the result of a collective and social process that, after debates that are often prolonged in time, end when a very large majority of experts converge or agree on a given theory or hypothesis. According to my claim, processes like these, which lead to the formation of a reasoned and consensual opinion play, but much more should play an analogous role in democratic institutions, especially when political decisions have important, widespread social consequences on issues like public health, industrial planning, climatic challenges, etc., and thus require a general agreement. The physicist Lee Smolin and the philosopher Roberto Mangabeira expressed the same concept:

> Scientific communities, and the larger democratic societies from which they evolved, progress because their work is governed by two basic principles.
>
> 1. When rational argument from public evidence suffices to decide a question, it must be so decided.
> 2. When rational argument from public evidence does not suffice to decide a question, the community must encourage a diverse range of viewpoints and hypotheses consistent with a good-faith attempt to develop convincing public evidence. I call these the principles of the open future.[5]

We must remember what was said earlier: scientific progress is often made possible by a free discussion that takes into account criticisms and objections to the dominant theories made by a minority of scientists or even, extremely rarely, by single geniuses. Einstein's 1905 revolutionary thesis on the dual nature of light was met with a certain opposition also among the experts in the field. Based on publicly available observational facts and theoretical hypotheses, that thesis were quickly accepted by the rest of the community of physicists and Einstein was awarded the Nobel Prize in Physics not for his theory of relativity but exactly for his revolutionary theory about the dual nature of light (wave-like and particle-like).

This fact, which is typical of the way in which scientific knowledge progresses, constitutes an important argument against the weakening of dissent also in modern democracies. This is why the most rational and fairest decisions taken in a democracy are the results of criticism of the minority, first approved in parliamentary commissions and then transformed into laws that are for the advantage of the whole community, thanks to a process at the end of which a compromise can be reached combining the most important common values.[6]

We have seen why the fallibility of science explains the change and dynamism of scientific knowledge and thus its progress. Since a theory so far corroborated in the future may not prove valid in every domain of its applications, it has often been the case that another theory that corrects the former has been proposed and then confirmed. Interestingly, both in science and in a democracy based on tolerance, a constructive criticism does not imply a complete "suppression" of the point of view of others. When a scientific revolution initiated by a minority of scientists conquers the consensus of a large majority and then of all the scientists, it typically considers ("respects") the preceding theories, at least whenever they happen to be a particular case of the more general, later ones. As we have seen in the two examples presented in the previous chapter (undulatory theory of light and gravitation), in these cases the preceding theories did not prove wrong in *all* areas of their application and for this reason they are still taught as part of science.[7]

As an additional example, in the very accurate and fundamental physical theory we rely on nowadays (quantum mechanics), the opposition of many scientists to formulations or interpretations alternative to the standard ones has already proved detrimental, given that the latter have suggested important experimental and technological developments (Chapter 2, n. 4). Such new formulations, some of which could soon have either experimental confirmation or falsification, reproduce the precise predictions of standard quantum mechanics but, more than that, they propose clear solutions to conceptual difficulties generated by the standard theory as it is usually taught in scientific textbooks. The same mechanism should lead to the formation of consensus in an ideal democracy: the sort of criticism that grows out from the minority should be encouraged and respected, in the same sense in which the criticism coming from those scientists who work in the so-called "boundary science" (in which the direction that the future research will take is not already clear) will need to be taken in due account.

3.3 Democracy as Separation of Powers and the Cross-control of Scientific Hypotheses

What I have schematically highlighted as the second essential feature of a democracy—the fact that the decision taken by the majority is the fundamental criterion which all citizens belonging to a democracy must bind themselves to—generates an important problem about the relationship between the role of the majority and the minority. Independently of its numerical consistency, the actual existence of a minority implies that an indistinct and homogenous notion of "the people"[8]—which many illustrious political philosophers defended, and on which many kinds of populism rely—is unjustified as well as potentially dangerous. The threat consists in the fact that the minority's duty and right to criticize the majority would be infringed upon. It must be noted that an analogous principle also applies to the choices about which single individuals are doubtful. Even though when we make a decision the totality of values that back it up is stronger, this does not mean that the motivations that were pushing us in the opposite direction do not exist at all and are not to be recognized after the decision has already been taken.

From this point of view, the French political philosopher Alexis de Tocqueville (1805–59), who visited the United States in 1831–2 and wrote an important book entitled *Democracy in America* (Tocqueville 1835–40), claimed that American democracy was subject to the risk of "the tyranny of the majority," and, hence to the possibility that the freedom of the individual to oppose the opinion of the state could be weakened or extinguished. Under the influence of Tocqueville, similar views were formulated by the already quoted John Stuart Mill in an important book entitled *On Freedom* ([1859] 2015).

The possible degenerations due to the occurrence of a tyranny of the majority could only be prevented by remembering the other—often overlooked—fundamental feature of a good democracy, that is, the separation or division of powers. We have seen how scientific hypotheses cannot be established by decree, but they need constant and accurate empirical control carried out by different groups of experts who work in the same field. Analogously, as is well known, the separation of powers, provided for in parliamentary democracy, has the fundamental aim to establish a system of "checks and balances" between the executive power (government), the legislative power (parliament), and the judiciary power. The democratic constitutions usually regulate how these three powers must ideally interact among them, without anyone prevailing over the other; what is otherwise at risk is the bad functioning or even the disappearance of democracy tout court.

The necessity of checking the government's activities (be this put in place by a single party or by more parties in coalition) is even more evident and can be further defended on the basis of *two* important principles that characterize the scientific methodology and that will be illustrated below:

1. According to Popper, the first problem in a democracy is not to answer the old question, "Who should rule?" ("The best ones," is the typical answer), but rather: "How can we best avoid situations in which a bad ruler causes too much harm?"[9] Note the analogy between falsificationism and his theory of government. In science, it must always be possible to specify conditions that, if observed, would be sufficient to reject a scientific hypothesis by hoping to replace it with another one. In the same sense, it must always be possible to "replace"—by means of free elections—a ruling party if its proposals to solve urgent "problems" failed. A physician must cure her patients with medicines that, above all, will not harm.[10] Analogously, a constitutional system should envisage norms that prevent the possibility that rulers "harm" the citizens and can do so for too long. For this purpose, to the already mentioned principles of, (i) pluralism, (ii) democratic alternation of parties and ensured free election, and (iii) the freedom of press opinion, teaching, religion, etc., one must add the reciprocal checks and balances of powers, which is equally important and related to all of them.

The analogy between the controllability of the prerogatives of powers and the decisions of political parties and the controllably of scientific hypotheses in the sense highlighted in the second chapter needs further assessment. By developing Popper's suggestion, one could claim that the social control of a scientific hypothesis by the communities of experts, together with the objectivity of the relationship between hypotheses and facts (as seen in the example of the rain given above) is the most efficient way to contrast the principle of authority, and hence to reduce the risk that a community of scientists, will "impose" its point of view on other communities without providing objective reasons to do so. Similarly, ruling political parties ideally should not approve or pass new laws without some form of contribution coming from the minority, that is, without giving the latter the possibility of examining and criticizing them. Before imposing them on all citizens, the ruling party ought to *justify* its legislative proposals by making possible, and even encouraging, a critical debate with the opposition and with the public opinion (including members of the party that voted for the government in charge).

The structure of the controllability of scientific hypotheses on the part of the scientific communities has to a certain extent similar features. The weakening of the legislative power to the detriment of the executive power undermines, at the same time, the role of critical discussion. However, direct democracies making room for too many and too frequent public discussions do not lead to decisions and thus to the resolution of urgent problems that the excutive power must make. This aspect of a good-functioning democracy is *partially* reflected by the fact that after a period of debates and discussions, by applying relevant epistemic norms the communities of scientists *decide* in favor of a theory and for objective reasons regard it as, at least temporarily, well confirmed. This fact does not contradict the importance of dissenting research programs, given the existence of areas in which the direction of future research is still not clear. From a certain moment onward, and until successive relevant evidence is brought to light, research programs leading to put in doubt the corroborated hypotheses are deemed sterile: analogously to what happens in high energy physics, sooner or later experiments must end (Galison 1987).

As is well known, in democracies, the judiciary power has the task to enforce the laws via appropriate sanctions, but it must relate at the same time to the other two powers by respecting their prerogatives. In a different but related sense, also in science there are actions and behaviors violating the norms of scientific research that are sanctioned by the community: impostors, authors of scientific frauds, defenders of hypotheses that they know to be wrong and plagiarists eventually, sometimes too late, lose their reputation. Furthermore, as in a free democracy, unless a scientific conjecture violates accepted ethical norms of the community (experimenting on live human beings), political censure must not intervene in the evaluation of scientific theories: there is neither an Arian, a Jewish, nor a bourgeois physics as has sometimes been advocated by ideologists of last century's totalitarianisms.

2. The second principle is an answer to a natural objection arising from the sort of consideration raised by (1) above: differently from what happens in scientific communities, in democracies there are plural interests and values that are regarded as irreconcilable among them. In politics, there cannot be any universal consensus because there are no facts that, as in the case of science,

can make us decide between rival political programs that are based on a conflict in their respective aims.

To answer this important and inescapable objection—which I must account for to highlight telling differences between the formation of consensus in science and democracy—I must further develop the claims stated in the Introduction, which constitute the second principle mentioned above: namely, that every theory or hypothesis, either scientific or more generally social, is an attempt to solve a problem (in the broadest sense of the term).

This conception is due to the pragmatism of Peirce and, in a different sense, to Dewey and in the last hundred year or so, has been developed by so many philosophers that listing their names would be an impossible task. Science begins with problems, that is, with conflicts between expectations belonging to our background knowledge and observations or experiments. In order to solve the problem, scientists put forth new hypotheses, the possible refutation of which "forces" us to try new hypotheses. The aim of scientific activity is therefore to solve problems more or less temporarily by finding a new agreement between new hypotheses and facts.

Analogously, it could be argued that the program proposed by a political party is a set of "hypotheses" that, at least in some cases, are aimed to solve social problems that are regarded as crucial by the vast majority, if not by all citizens. Think, for example, of a high unemployment rate, the diffusion of a pandemic, uncollected rubbish on the streets, the pollution of the oceans, stinking municipal sewage, public services of bad-quality transport, health, etc.[11] Clearly, in their words and hence in their programs, in order to win the elections, every party claims that it wants to fight against these phenomena. The fundamental point, as the Italian political philosopher Norberto Bobbio reminded us, is, however, that all citizens must be informed in a precise way on *how* the political parties intend to realize such aims! With what instruments? For instance, by adopting which fiscal and economic policy?

We can reasonably suppose that among the choices available to solve problems of common interest, there are some that are objectively more effective than others and that such efficacy does not necessarily infringe on other fundamental values, first and foremost the equality of citizens before the law, their dignity, their rights to healthcare, etc. The conflict between political parties of different political orientation can then be based on *objective facts* that are shared and ought to be shared by everyone and that are investigated by experts whose different political orientation, at least in theory, does not have any influence on the effectiveness of the suggested solutions. As specified above, the experts[12] have the task to indicate objectively valid causal strategies (Cartwright 1983) to reach aims (e.g., to decrease unemployment) by means of "technical tools" (raising taxes, reducing public expenses, etc.) on which there can be and should be the same kind of debate that features in reaching consensus in scientific communities.

In fact, once some *common* aims have been established—something that in a democracy is not always possible but often possible only for subgroups of the society—the most appropriate tools to reach those aims are typically based on observations, statistical data, etc. These means–ends relationships are operative also in those everyday circumstances in which, for example, when we want to go home (our aim), we think whether to take the metro, the car, or the bus, and decide among these three alternatives (our tools) on the basis of our personal experience, as well as by taking into account other factors such as costs, comfort, and other people's experience.

Although the modality of choice of the tools to reach a political objective of common interest is very different and much more complex than that typically employed in the scientific inquiry, the reasoning at its root is the same. The indispensability of the intervention of experts for the actual reaching of a social objective, independently of whether it is universally shared, is provided by the fact that the relationship between means and aims is *causal*. Whenever I carry out X (the means) and I systematically obtain Y (the aim), if nothing prevents the regular succession of events, I can conclude that "X is the cause of Y."[13]

It is important to note that such a causal relationship plays an important role both in the natural and social sciences. In the latter, the means chosen to solve a certain problem are linked to other values: raising the public deficit to limit the problem of unemployment has some predictable social consequences that not all citizens regard as desirable. But note two important points. On the one hand, it is the reliability of the means–ends relationship that brings to light these disagreements in values. On the other, a pluralistic variety of viewpoints intervenes because conflicts of values are made more acute by the fact that predictions in the economic or social areas are much more uncertain than those allowed for by the physical laws. Therefore, the difference among the plans of the different parties can depend both on their different values and on the means that should be adopted to reach some common value. In the natural sciences, on the contrary, the difference in aims is *less* significant: despite the existence of different scientific styles, experimental accuracy, simplicity, unification, and explanatory power are shared epistemic values of the scientific enterprise (see Kuhn 1977). However, also in the social sciences, open discussions about which methods are more effective in reaching shared aims can induce the citizens to harmonize their different values and get to a better solution.

The prisoner dilemma illustrates rather clearly that cooperation among individuals is often more advantageous than conflict. To maximize their profit, two prisoners who cannot communicate with each other are each tempted to accuse or betray the other. The game is conceived in such a way that if one of the two cooperates and the other betrays, the profit for the betrayer would be maximum (say, she would be acquitted and the other would be imprisoned for four years). If they decided to cooperate, each would be acquitted after four months. To maximize their profit, both are tempted to betray the other but, in this case, the loss would be greatest for both (both four years of prison). This prisoner dilemma shows that in many situations, deciding not to betray the other (i.e., cooperating with the other prisoner) benefits both prisoners more than it would have if each had yielded to the temptation to betray the other. Our societies must be organized in such a way that the "free rider" who wants to maximize her profit by not paying the ticket paid by the others must be relatively certain that she will eventually be worse off for purely selfish reasons.

Summing up, if the strategies adopted by a political party and aimed to solve a shared social problem end up being ineffective, the political party has failed in its mission. The defense of the principle of political turnover, which constitutes a pillar of democracy, is mirrored and finds an important justification in the method of trial and error, which is typical of science. If a political party in power could not be substituted, it would be as if a scientific theory that does not work properly could not be replaced. The consequence in the first case is a dictatorship, in the second a sterile dogmatism. In periods of theoretical changes in science, a variety of hypotheses (and even, more controversially, methods) to be further pursued makes the solution of a new problem more probable (Feyerabend 2010). In a democracy, we need more parties from which to choose, following the fact that the party in power may be unsuccessful in solving problems of common interest, thereby failing (even if perhaps only temporarily) its reformer mission. Therefore, the value of pluralism and, hence, of tolerance of other people's opinions is a necessary condition for the good functioning of both science and democracy.

4

Representative Democracy, Direct Democracy, and Scientific Specialization

4.1 Two Kinds of Democracies

In the previous chapter, I compared the relationship between the main principles of democracy (equality, majority, and the separation of powers) with those characterizing the social structure of science. In this chapter, I shall tackle a distinction within the concept of democracy by showing that it is strictly related to the values that make the progress of scientific knowledge possible. Such a distinction concerns a more controversial topic, that is, the principle of representativity and, hence, the representational or direct nature of democracy.

Schematically, the two types of institutions discussed in this chapter can be characterized as "democratic," given that in both forms of government the three features described in the previous chapter apply. However, as will be shown, in direct democracies, the value of pluralism, which is fundamental for both science and democracy in the three main senses considered above, is much more at risk. *Prima facie*, a representative democracy is based on a mandate of our decisional power transferred to some delegates chosen by means of a political vote. One of the most important justifications for this principle is that the delegates are more competent than their electors in tackling problems whose solution needs specialized knowledge. Conversely, direct democracies are based on the thesis that citizens must be able to express their opinion much more directly on the important questions that concern them, in such a way as to reduce the mediation on the part of delegates to what is strictly necessary.

By recalling the key methodological point introduced at the beginning of the first chapter, also in what follows the two kinds of democracy that I will be discussing are two *ideal types* (*Idealtypen*). That is, also representative democracies allow for some direct forms of expression of common will, such as the referendum and, analogously, also direct democracies can make room for institutions that take decisions for the citizens in an indirect way. The remaining, important distinction between these two forms of democracy relates to the values that ground the decisional procedures of science in comparison to those characterizing direct and representative democracies.

4.2 Populism, Direct Democracy, and its Problems

I use the label "direct democracy" to refer to forms of government that can be implemented in different ways, but that share a common criticism towards the typical forms of representative democracy. Despite the ideal-typical character of the dissimilarity, the main features of the latter are in sharp contrast with the ideals of the former, which were historically implemented in the public space of the agora, the main square in the cities of ancient Greece. In these places, all free (non-slaves) male and well-to-do citizens gathered to deliberate directly on problems of common interest such as laws, war and peace, or the ten-year banning of an individual from the community (the so-called "ostracism"). There were no elected delegates who decided on their behalf. After some political phenomena that we could refer to as "direct democracy" and that was implemented in Italian cities during the early Middle Ages, historians usually define other, ephemeral experiments of direct democracy as, (i) the Jacobin phase of the French Revolution (1792–3), (ii) the Paris Commune in 1871, and (iii) the attempt carried out by the "Soviet of farmers and workers" during the Russian Revolution (the attempt being extinguished when Stalin took power).

Before discussing the problematic aspects of direct democracies, it is better to highlight three aspects that could be regarded as "positive" and that are usually glorified by its supporters. All of them involve the ideal of autonomous choice. *First*, as the most authoritative proposer of direct democracy in modern times, the Swiss philosopher Rousseau argued, unlike the electoral, representative systems, direct democracies[1] implement an ideal of *moral autonomy* since no one decides in the place of the individual citizen. Given that nobody doubts that freedom essentially consists in the exercise of our own capabilities to decide, for Rousseau, paradoxically, in representative democracies the only moment in which the citizen is truly free is when she casts a vote. According to him, freedom for the English people—who at the time lived under a regime that we could call with some anachronism "representative democracy"—boiled down to this.[2] In the lapse of time separating one election from the next, Rousseau claimed that the English citizens were nothing but "moral slaves" who thought only illusorily of themselves as being free.

The *second* positive aspect is related to the first: in direct democracies, citizens can look after their own interests better than anybody else to whom they gave the mandate to act on their behalf. By not delegating others, they are fully responsible for their political decisions. *Third*, direct, participatory democracy should in theory develop a stronger sense of belonging to a wide *community*, in which the citizen can discuss freely all options and then decide all together. On the contrary, representative democracies tend to isolate and alienate the citizen from the values of the society she belongs to, thereby making her social role completely anonymous, a fact that is symbolized by the rule that the voters cast their vote into the secret ballot box.

But how can one presume that there is no need to "negotiate" between different types of interests and values, a mediating role that in the representative systems is usually played by the elected? Typically, the defenders of direct democracies respond to this question by adopting a distinctive conception of human beings, since they hold that the individuals' fundamental interests and values are very similar among them, if not, identical. In this way, ideally, the *will* of every individual can freely coincide with the *common will* (Rousseau's term), so that everyone can think of herself as being free *because* her will coincides with that of others. In his *Social Contract* (1762), Rousseau's common will is regarded as a collective entity representing all citizens. As such, it is something *more* than the sum of the single individual's will, in the same way in which a concert is something more than the sum of the single notes composing it or of the voices that each instrument provides to the whole orchestra.

In this sense, the will is "common" because it is aimed at the common good which does not coincide with the will of the majority. Analogously to the concept of "the people," already mentioned before, the common will does not present at its core any significant differentiation in terms of values or interests and, hence, no important forms of pluralism or dissent. I already explained why, with respect to science, these two values are indispensable for the growth of knowledge. In a few words, from the viewpoint of a direct democracy, the freedom of individuals does not consist in pursuing individual ideals of life (hence, potentially different, and pluralistic) but in the direct participation of citizens in sharing crucial public decisions.

Moving now onto the rather problematic aspects characterizing the models of direct democracies, I shall briefly deal with four of them which should convince the reader why representative democracies are not only preferable but necessary.

The *first* argument against direct democracies relates to their probable degeneration into authoritarian forms of power. The complexity of this topic, which here cannot be discussed in full, can nevertheless be forcefully summarized with the help of the following quotation: "Rousseau's individual, requested from dawn to dusk to participate to common discussions in order to exercise her political duties as a citizen would be not the 'total' man but the 'total' citizen ... and the latter is the other face, not less threatening, of the total State" (Bobbio 1987: 41). The reason for this claim is easily stated: the complete identification of the private with the public sphere puts at risk the freedom of individuals to pursue interests and aims that, despite the existence of overarching, shared values, do not necessarily coincide with those of others.

The *second* argument that can be brought to bear in favor of representative democracies was suggested by a philosopher and writer of the generation after Rousseau's, namely Benjamin Constant (1767–1830). In his talk *The Liberty of the Ancients Compared to that of Moderns* delivered in Paris in 1819, Constant claimed that in the already crowded European cities, which were organized in a much more complex way than the small cities of ancient Greece, the ideal that all citizens participate in all the public decisions is unattainable (Constant 1819).[3] Considering the exponential growth of the world population, after two hundred years, Constant's argument is even more compelling; it is not surprising that forms of direct democracy are more widely spread in small countries with a low population, such as Switzerland. Moreover, in complex contemporary societies, the number of topics of public interest is certainly much higher than the issues that were important for the ancient Greeks.

Third, to strengthen this purely numerical argument, one must take into due account the fact that in the Greek world, the possibility to deliberate in a collectively and directly way on common interest topics, according to Constant, implied a slavery economy that could well allow free male citizens to have enough time to take part in assemblies. Everyone who must work to live, as happens in contemporary societies, is forced to delegate someone else to work full-time to defend her interests. According to Constant, who wrote after the French Revolution, the commercial needs of the bourgeois society made an indirect or representative democracy indispensable. The main aim of this kind of institution is to protect everyone's right to pursue autonomous ideals of life, which, pace Rousseau, do not necessarily coincide with those of others. It must be admitted that in Switzerland, citizens who are very busy with their private enterprises are often called to vote by means of public referendum.[4] The average level of wealth in this country, however, is very high and economic conflicts (that usually require important political mediation) are much less numerous than anywhere else.

It could seem that the last two objections, based on the limited amount of time that numerous citizens can dedicate to reach collective decisions, can be overcome by relying on digital technologies. Is it unthinkable that a very large community of citizens can deliberate directly and quickly on important questions by means of a vote cast via the internet, thus avoiding the slowness of decisions taken by our representatives?

In addition to the remarkable technical difficulties that would be necessary to avoid manipulations of votes cast via the internet, the *fourth* argument against direct forms of democracy—whether they are realized with the web or not—is as strong as the others. Even though the web can play an important role in the formation of political, social democracy-oriented organizations, the growth of specialist knowledge, both in the natural and social sciences, calls for diverse forms of indispensable political mediation.

It is important to stress the oft-neglected contrast between the source of scientific knowledge and the nature of the enormous quantity of information traveling via the web. Unlike the latter, scientific knowledge accumulates slowly, has many different sources of validation, and is characterized by many links among its different critical components. We have seen why in science there is neither a single nor few sources of epistemic authorities: the authority depends on the social nature of science illustrated in the second chapter.

The scientific web of beliefs is highly complex since its nodes are very interconnected, and this social control facilitates the elimination of errors, frauds, or fakes. Scientific knowledge is complex in this precise sense and interconnectedness is a typical feature of all complex systems (our brain included). For instance, storms of starlings, which are remarkably complex systems, form amazing patterns in the sky because there is no leader that each bird follows:[5] each bird is guided by a few starlings nearby and this fact enhances the chances of survival of the storm. On the contrary, even if connected with a *potentially* enormous amount of information available on the web, the *sources* of reliable data in the internet is negligible and scant if compared with the immense network of validated scientific hypotheses.[6] It frequently happens that a few links *dominate* all the others, so that the users of the latter uncritically follow the former. This fact has important consequences on the way information propagates. The sources of scientific and political information traveling via the social networks do not undergo the same number of controls that science is subject to. In other words, in the web, errors and fake news spread faster and are very difficult to filter because the number of nodes or links controlled by others is negligible. In the metaphor adopted above, on the internet there is only one starling or a few dominating starlings which all the others ineffectively follow.

Consequently, in some important respects, the very idea of equality that is at the root of democracy and that in the forms of direct democracy is even more highlighted and sought after, is in conflict with the indisputable fact that as a consequence of the division of knowledge, we are *not* equal.[7] Physicists dealing with the origin of galaxies are familiar with a branch of physics that is largely inaccessible not only to a learned readership (biologists, psychologists, economists, sociologists, and so on) but also to other physicists specialized in other branches of the same discipline. This fact is well illustrated by the authoritative historian and philosopher of science, Thomas Kuhn, who stated that: "although it has become customary, and is surely proper, to deplore the widening gulf that separates the professional scientist from his colleagues in other fields, too little attention is paid to the essential relationship between that gulf and the mechanisms intrinsic to scientific advance" (Kuhn [1962] 2012: 53).

The specialization of scientific knowledge seems to be almost inevitably pushing toward representative forms of democracy and to provide, at the same time, the strongest reasons behind it. The entrustment of our beliefs to experts working in different fields corresponds to the necessity of entrusting the realization of our ideals to those we elect on the basis of the principle of competence. A feature of our rational choices is that the quantity and quality of information need to be as high as possible. There are technical issues (say, the harmfulness of the presence of mercury in fishes, the consequence of the presence of estrogens in cows, or climate change due to ozone-layer depletion, etc.) of immense social consequences on which the citizen must decide that, however, presuppose a reliable information that can be provided only by real experts or mediators of knowledge.

It is very difficult to avoid the conclusion that inexperienced citizens cannot directly deliberate via the web (as in a cyber democracy) on problems like these; and surely, they should not do that before being duly informed by competent people of the consequences of their choices. Nevertheless, how can we find a way to convince the web user of the fact that the opinion of a scientist or a group of scientists does not have the same value as that of a single individual or of a group of non-specialists, in particular if the opinions diverge?

4.3 The Risk of Technocracy: Three Objections to the Compatibility of Autonomous Decisions and Delegations to Experts

A possible answer to the question just stated contains a potential double-sided weapon that can be easily wielded by followers of direct democracies.

On the one hand, in a time dominated by technology, there is the risk that representative democracy degenerates into a technocracy, i.e., as defined above, government composed only of technicians in which the decisional autonomy of necessarily uninformed citizens is impoverished and is wholly transferred to the expert deciding on her own behalf.[8] On the other hand, the necessary solution to the fundamental conflict between the principle of representation required by the division of knowledge and the principle of citizens' decisional autonomy cannot consist in having everybody express their opinions on every subject.

In other words, the challenging task ahead consists in organizing representative democracies in such a way that the rationality of a decision based on the principle of competence and representation can coexist with our autonomy of choice, which is only deceptively made possible by direct forms of democracies. The balance between these two requirements is extremely difficult to reach because they pull in opposite directions; it is an ideal towards which we must aim, although in representative democracies it has already been partially implemented.

Since the critical discussion between two or more contrasting positions is the best way to have the most reasonable thesis prevail, I must examine the possible objections to what has been said so far in favor of representative democracies by using the model of a trial, where both the prosecutor and defender try to convince the jurors of their favored thesis. More in line with what has been said so far, an examination of the different arguments in favor or against one of the two institutional models can be likened to a debate between two scientific hypotheses that have not yet achieved consensus among the experts. In our case, the best defense of the principles of representative democracy will go through two argumentative steps. First, I shall discuss *three* serious objections to the reconcilability of the citizens' principle of autonomy alongside the principle of delegation, and then, second, I shall refute them all, one by one. I shall thus follow the dialectical strategy that Galileo (1564–1642) adopted in his dialogues and in the trials in which he was prosecuted. He reinforced his opposer's thesis in the strongest possible way so that the ensuing refutation made his own viewpoint even more convincing. It is easy to prevail in a discussion if your opponent's views are read uncharitably and presented unfairly as absurd!

Each of the three arguments that I am about to present presupposes that also the defenders of representative democracy, like their opponents, have in the utmost consideration the importance of the value of autonomy of choice. Unlike what happens in a paternalistic and intrusive state, in every real democracy the citizens must in some way make their autonomous opinion heard on everything important concerning them. What separates direct democracy supporters from those in favor of representative democracy is *how* they want to implement such an ideal in an efficient way. The following claims, if sound, would become decisive objections *against* the desirability of representative democracies, insofar as it is motivated by the necessity of epistemological mediation deriving from the principle of competence. As we shall see, however, each of these arguments can be rebutted in a way that I regard as decisive. Moreover, as we shall see, the ensuing replies against the supporter of direct democracies will give me the opportunity to shed more light on the key value of the public controllability of science already broached in Chapter 2.

By turning the weapon of the theorists of representative democracy against themselves, the first objection against the reconcilability of two principles of competence and autonomy surprisingly relies on the oft-mentioned, growing hyper-specialization of scientific knowledge. Recall Lippmann's three-layered model: the experts, the politicians who seek advice from the former, and "the phantom public." Even a politician with a thorough knowledge of economy, fiscal policies, environmental sustainability, and so on, that is, an expert elected by the citizens to make their interests fully accounted for, is always in need of additional advice on the part of super-experts who are *more* competent than her. In this case, however, the citizen's mandate to the politicians would become even more indirect. It follows that every solution to the problem of vouching for an autonomous choice in representative democracy would not only seem utopian, but would also bring the risk that the ultimate aims of political action, which in the last instance should be dictated by us all, scatter in a thousand creeks, without ever materializing.

Also, the second objection against the possibility that this conflict can be resolved within representative democracies relies on the same "weapon," namely the undeniable, growing specialization of scientific knowledge. How can the principle illustrated in Chapter 2 be justified, according to which the controllability of the public knowledge is the most solid ground on which a democracy lies? The very great majority of citizens cannot even in principle acquire all the necessary information to check a scientific hypothesis, say on the origin of the galaxies, on the bone structures of Neanderthal man, or on the function of the visual cortex of the human brain. Even the repeatability of an experiment (except in some domains that are not-experiment-based, as astronomy), which was considered in the second chapter as an essential value of the public and non-esoteric features of science, is often realized by very small groups of experts whose members have very different competencies and, hence, not by a single person. The most striking example is provided by the experiments carried out in Geneva for the discovery of the Higgs boson, which implies the collaboration of thousands of physicists, mathematicians, engineers, and other scientists. These experiments, that accelerate particles to a speed close to that of light, are very *expensive*, and need a financial collaboration among many countries, a fact that prevents scientists from repeating them "too often" (Galison 1987). Lacking similar accelerators they cannot be repeated in other regions of the earth, even though this fact does not detract from their experimental reliability.

This kind of "big science" (Galison and Hevly 1992), which is mostly a consequence of the Second World War Los Alamos project, is not even comparable to the experiments that were carried out in 1672 by Newton who, alone in his laboratory, and using some prisms, managed to show that white light has a composite nature (Newton 1671-2). It is only *these* sorts of experiments that can be repeated by all individuals equipped with similar instruments, but they do not play an important role in the validation of contemporary physical theories.

In view of all of this, which concerns the repeatability and, hence, the controllability of scientific hypotheses, it seems to follow that the principle of citizens' autonomous choice is *incompatible* with the growing delegation to experts on which the defender of representative democracy insists. All these facts seem to imply the necessity that political decisions be, to put it vaguely, somewhat less mediated in ways that should be studied more: the decisional autonomy of the single individuals would decrease the risk of paternalism, that is, the risk of a "state father" that, by means of experts, tells citizens how to behave so that they do not harm themselves, as children sometimes do.

To illustrate the fault of paternalism, we could refer—with an amusing example—to a section taken from an important and very influential essay on the nature of illuminism that Kant (1724-1801) wrote in 1784. In this essay, Kant remarked that no one's diet should be decided by a doctor (an expert, i.e., a dietologist) that is, by someone who decides on our behalf what we should eat:

> If I have a book that thinks for me, a spiritual guide who has conscience on my behalf, a *physician who decides the diet that is most suitable for me*, etc., I don't need to bother myself. I have no need to think on my own if I can only pay; others will already take over the annoying business for me. I no longer need to think on my own. As far as I can pay for it, I do not need to think

anymore: someone else will have the responsibility to carry this boring task out on my behalf.

Kant 1784; my emphasis⁹

These kinds of heteronomy (which consists in obeying rules established by others) deny the very ideal of autonomy (e.g., consisting in giving rules to oneself to decide how to act) on which Kant based his whole moral philosophy. If Kant, who thought of the intellectual and ethical autonomy as the greatest value of our life, had seen a rich, overweight Western person (who does not know the bite of hunger) entering the office of a nutritionist,[10] he would have concluded with Rousseau that we are condemning ourselves to live in a state of constant slavery or minority. Even more, to use once again a metaphor from Kant, we would live as children held by the braces by someone who helps us walk even when we should not need it anymore. After repeatedly falling, we would learn to walk without any help from our parents, and only parents can know which kind of food must be fed to their children.

Given the importance of the ideal of intellectual and moral autonomy, this argument against the necessity of political intermediation seems very strong.

4.4 The Confutation of the Three Objections and the Social Character of Knowledge

To argue against these three arguments, let us start with the last one, which is easier to refute. Even though Kant's example about the physician who tells us what to eat is no more than a provocative argument to ridicule those who refuse to be intellectually and practically autonomous, it allows me to introduce the argument that I want to develop. In Kant's time, medicine was obviously less developed: blood tests reporting the exact levels of cholesterol or glucose were unavailable. Today, not only have GPs but also "all-purpose" orthopedics left their places—with occasionally negative consequences—to the orthopedic who is specialized in treating knees, feet, spine, fingers, etc., but, in addition, phenomena like these seem to be irreversible.

We have thus to give up, qua unobtainable, the kind of autonomy teasingly exemplified in Kant's quotation. At the same time, we have to show, as I shall do in the next sections, that the indispensable intervention of experts does not prevent free choices on the part of the citizen living in a representative democracy. This claim can be illustrated by means of two examples taken from ordinary life, which will also be useful to tackle the problem of trust in the expert, which I shall examine more extensively in the next chapter.

The first example is that of a very competent lawyer whom we trust since we consider her to be the best defender of our legal rights. On the ground of all the relevant facts that we communicate to her, we must trust the lawyer almost blindly and rely on those strategies she chooses in order to reach *our aims*—which clearly depends on us. This reliance has a decisive value, even if in the last instance it is the lawyer who decides the best way forward.

In spite of the fact that it is seemingly more dubious, the second example illustrates the same conclusion. Our decisional autonomy seems to be more threatened when we call for a doctor, since the latter can decide on questions relating to life and death. A doctor's knowledge is, however, an indispensable tool to regain health or prevent future disease. Only a doctor or a group of doctors can provide us with reliable information on, for example, the chances of recovery without surgery, the collateral dangers of surgery, and so forth. However, even if *it is our aim* to become healthy again, how could we defend the claim that the decision of undergoing surgery is only up to us, given that it has been recommended by an expert?

If we need to extract a painful tooth, the decision to trust the dentist does not seem to harm our freedom. In much more complex cases, which involve risky surgeries, however, we must sign a document for informed consent. Clearly, this document is also useful for the surgeon and for the hospital to decline any responsibility for the outcomes of the recommended surgery. In this sense, it is irrelevant with respect to our freedom of choice. However, also this example is useful to illustrate my point. Informed consent exemplifies the ideal of decisional autonomy that according to the theorists of direct democracy would be endangered by the presence of the expert: we sign the document because we agree on the surgery by accepting its potential risks.

Although one can lessen the importance of informed consent for the defense of the autonomy of the individual choice, there is an important argument to be taken into due account. Such an argument, while calling into question the aggressive medical treatment, reinforces my point of view on the full reconcilability between the two principles exactly under those circumstances most concerning our dignity as human beings. To a patient who will very probably need an artificial ventilator for the rest of her life, the possibility to decide autonomously on the surgery must be given. Such a decision is autonomous, and at the same time based on scientific evidence. The living wills are necessary for vouching for the individual's free decision—before she is unable to do so–to accept an operation whose consequences are scientifically foreseeable. Such wills enable the individual to decide autonomously on the basis of what we know about the consequences of the operation.

With all the necessary distinctions, and despite their simplicity, these two examples can be generalized and applied to political decisions made by the collectivity. The aspect to be highlighted at this point is that the rationality of any decision—and hence also of those taking place in institutions that respect the individual citizen's choices—lies in the already mentioned relationship between means and ends that at this point needs further development. Both the lawyer's client and the doctor's patient pursue individual aims that these two experts must implement by relying on their professional knowledge. This entails that they must choose the best means to safeguard the individual interests of those who called for them. Analogously, in collective decisions, the experts' indispensable specialized knowledge is necessary to both citizens and politicians to find out the most suitable ways to implement autonomously chosen aims.

By using a metaphor employed by sociologist Max Weber, I can highlight this principle through a very effective metaphor: the task of a scientist or of an expert is to provide us with a map that helps us decide *how* to go where *we want to go*. We decide the destination of our trip, together with the persons elected by us, including experts. That is, qua experts, they are called to suggest the most effective means that reach the ends of the rest of the citizens; qua voters, they are entitled to exert their autonomy and choose their representatives as any other citizen without telling her what to do. Despite the inevitable limits due to the unequal division of knowledge and to specialization, the principle of the delegation to experts and politicians who act on our behalf warrants both the fact that the strategies implemented to reach such aims are the most effective[11] and the fact that important decisions are taken autonomously.

To rebut the second objection—which insisted on the impossibility to control the experts' knowledge even in principle—I must expand what I wrote in the second chapter with two additional arguments:

1. Above all, it is worth providing some explanation for the expression "lack of controllability in principle." Even if neither time nor the epistemological skills of a single individual or of a group of scientists suffice for a direct control of a single hypothesis, it is not such a type of control that really matters. When I used the expression "control in principle," I was referring to a suitable process of trust in the experts and not to the possibility of personally carrying out the control of any hypothesis. The fundamental point concerns the social procedures through which our own beliefs are justified, which perfectly mirrors what happens when a scientific hypothesis put forth by a group of scientists is accepted as reliable by scientists or experts working in very different and remote fields of inquiry.

The objection about the impossibility of "controlling in principle" loses its value if we think that most scientific knowledge *already acquired* does not need to undergo a further direct control on the part of the scientific community (and, hence, of the single individual) to consider it as *reliable*. As already said, scientists do not doubt every theory, nor do they believe in all theories: calling a set of beliefs into question implies the existence of another large set being taken as epistemologically reliable. Only the areas in which scientific knowledge is still uncertain and has not reached any consensus are put under critical scrutiny by scientific communities.

Like any scientist, citizens who trust the hypotheses taken for granted by scientific communities do not need to personally check all of them. The *indirect* feature of their control—what I named "control in principle"—is grounded on the rationality of the epistemological mandate to a scientific community composed by experts who agree on the reliability of a given theory.

2. The second consideration, also aimed at the criticism exposed above, justifies, both for the individual and the scientific collectivity, the expression "controllability in principle" by grounding it on the assumption that a system of the reliable transmission of beliefs really exists. Independently of scientific

hyper-specialization, citizens and scientists can rationally trust the beliefs gathered by other scientific communities because the very great majority of such beliefs has been widely confirmed through the same methods and procedures for at least four centuries.

The controllability that the citizen can theoretically carry out, despite being very indirect, is epistemologically reliable because all human knowledge, including that on which we ground our daily behavior, has an inevitable social nature. Most of our everyday beliefs are based on those of others unless we have well-founded reasons to call them into question. The only important difference between common-sense beliefs and scientific hypotheses about a subject S is that when the latter are concerned, the epistemological mandate undergoes a much stricter and rigorous check, based on the cross-control of many experts in S. Even in cases of conflict between two persons holding opposite beliefs concerning *everyday life facts*, we can rarely personally check how things are or what it is "the truth." Often, we ask someone else whom we trust to dissolve our doubt so that, even in these cases, there can be only an indirect form of control.

These considerations entail that both in the case of science and in everyday life *only* indirect forms of control are possible and available. The key point is that also collective but "autonomous" decisions that are typically encouraged in direct democracies would in any case be grounded on the opinions of others. The difference with representative democracies based on the principle of competence depends on the fact that in the case of conflict of opinions, some "people" ought to be trusted more than others!

Therefore, the fact that the control of scientific hypotheses can only be indirect is decisive to refute the second objection put forward by the theorist of the epistemological and political dis-intermediation that was broached above. The assent we provide as single individuals and as members of a society to the credibility of the hypotheses put forward by a scientific community—a sort of mandate or mediation—is the result of a rational choice. On the condition that we become familiar with the general method by means of scientific hypotheses are tested and justified, such a choice is not only rational, but also free, and autonomous. As we shall see in the last chapter, familiarity with such a method is indispensable for the acquisition of a scientific mindset to problem-solving and is important regardless of the specific way in which a particular method of acquisition of knowledge is used in the different sciences. In sum, the inevitability of an indirect control of our scientific knowledge, if rational, justifies models of representative democracy in which competent politicians are responsible for finding out the most appropriate ways to reach our aims and guarantee the public interest.

Once the role of experts is established, however, citizens must turn their attention to those crucial cases in which there is dissent between experts, and between experts and pseudo-experts about hypotheses having relevant social consequences but that, in the latter case in particular, are *artificially* presented as subject to genuine controversies: what should we do in those crucial cases?

4.5 The Disagreement Among Experts

We should first discuss the kinds of conflicts that occur among scientists belonging to different camps, i.e. among honest and dishonest experts who intentionally diffuse false beliefs for economic reasons. Later I will discuss dissent involving pseudo-experts.

In the former case, which are the most difficult to handle, competent scientists themselves can produce fake news (Kourany and Carrier 2020) and we can never be sure that the final reports produced by panels being nominated by companies keen on the production of certain artefacts are as reliable as those produced by a panel made up of honest and competent experts. In view of scientific hyper-specialist knowledge, how is it possible for a citizen to choose between hypotheses advanced by two conflicting groups *of experts* if a possible *third* opinion expressed by another panel that should act as an independent judge could be partial as well as unreliable? It could happen that the panels that should act on the basis on the principle of independence are only useful in delaying decisions that should be taken rapidly (Oreskes and Conway 2010). In any case, the problem of trust towards the third panel of experts will apply again also vis-à-vis other panels.

Moreover, from the necessity to delegate to experts the task to provide us with instruments to decide as autonomously as possible, another fundamental problem originates. To the non-specialized citizen who does not perceive that unanimity of opinions that she expects from the experts (independently of whether the conflict involves pseudo-experts) it becomes natural, even if erroneous, to conclude that the lack of agreement on the part of those who are equally perceived as experts derives in any case from the fact that none of them is truly unbiased but rather serve different economic interests. Consequently, scientific knowledge loses any authority whatsoever, and is regarded as mere rhetoric means by an elite enrolled by powerful economic interest, with the dangers that one can imagine about the decision to take vaccines or measures against climate change.

The possibility to provide the citizen with reliable instruments that allow her to evaluate which one of two opposite views is more reliable is an indispensable condition for the actual survival of every democracy, which is thesis T_2 presented in the Introduction. It is only if we know how to choose the "right expert" or more plausibly the "right group" of experts that can we reconcile the autonomy of our decisions (Rousseau's ideal of direct democracy) with the principle of competence that I invoked to justify representative forms of democracy that are made necessary by the growing specialization of contemporary sciences.[12] If strategies that are strong enough to confer such a trust in a rational way did not exist, we would be condemned to live in a society that is subject to the risk of manipulation caused by scientific disinformation, that, as we well know today, is very widespread also due to the internet. The challenge ahead of us is therefore to provide the means enabling us to make our choices in a rational, autonomous, and informed way, particularly when the opinions do not look to be unanimous because of the incompetence or dishonesty of one of the two acting parties, or simply because the evidence is not sufficient. This is the topic of the next chapter.

5

Scientific Disinformation and the Distrust in Experts

Starting from the two simple examples of the doctor and the lawyer, in the previous chapter I have shown why trusting experts is a necessary condition to reach our ends and hence reconcile our decisional autonomy with the principle of competence. We have also seen why the latter is necessary to reach collective decisions and thus represents a decisive argument for privileging forms and models of representative democracy.

In this chapter, I shall first put forward three main reasons leading us to trust science and experts. Afterwards, I shall explain how such a trustworthy attitude makes it important to inquire into the main causes of the widespread perception that we are manipulated by scientists. Among these causes, the presence in the public debate of a *conflict of opinions* between real experts and self-proclaimed ones prevails; this divergence is intentionally created (as well as fostered) by scientific disinformation. I shall end the chapter with some necessary considerations that will not make the solutions I proposed too shallow. This implies a discussion of the problem of scientific frauds alongside the more delicate question being raised by the so-called *inductive risk*. The conclusion I shall reach is that scientific disinformation has two social consequences connected by a causal relation: doubts about the existence of an objective point of view prevailing over the scientist's individual interests generate an epistemological mistrust of science and of the beliefs of experts. The second consequence is that the same distrust also leads unreflexively to the abandonment of any idea of political delegation and, hence, to fall into the arms of direct democracy.

5.1 Three Reasons Why we Intuitively Trust Science

There is a strange contrast between the reasons making citizens doubt about the impartiality of scientists and experts and three intuitive beliefs that generate in all of us a well-placed trust in the validity of scientific knowledge and in those who work hard to implement it.

The first belief has a purely applicative and technological nature and is perceived by everybody as being true, independently of the scientists' political orientation. The second belief depends on the intersubjectivity of scientific knowledge, which is reflected in the already mentioned scientific community's consensus about a certain set of hypotheses, and which I will now approach from a psychological point of view. The third belief is linked to a philosophical thesis called "scientific realism": even if in the literature dealing with the science–democracy relationship this aspect is often neglected, it is the most important among the three because it grounds the first two and allows us to explain them in full.

Starting from the first aspect, no one doubts that *science works*. In addition to the indirect nature of the control of scientific beliefs, it must be stressed that under certain circumstances, we check the applicative consequences of many scientific theories *personally*, without knowing any of their theoretical aspects. This happens also when such theories are very far from our experience, as in the widely accepted fundamental theory of the physical world, i.e., quantum mechanics. We would not undergo eye surgery with a laser or an exam with a Positron Emission Tomography (PET), unless we were sure that the risks we run for our health are very low. The same applies to the laws of hydrodynamics, which we personally check when we board a boat or a flight.

It follows that for practical purposes each of us trusts completely, albeit in an unreflective manner, the kind of knowledge elaborated by experts in quantum mechanics or hydrodynamics, whose theories have made those technological applications possible. If the theories elaborated by the experts were unreliable, the technological applications of such theories would not be either. We trust the applicative aspect of science since we experiment with its practical consequences *directly* every day: implicitly, we trust scientists who with their theories have made such applications possible.

The second aspect leading us to believe in the epistemological effectiveness of science and in its experts' reliability is our perception of their consensus. In this case, our degree of certainty of the correctness of their hypotheses and of their trustworthiness is proportionally strengthened. This fact is easily explicable also from a psychological point of view and is rationally justifiable: when different witnesses who do not have serious interests to lie and have not interacted with one another report the same version to a judge, we spontaneously but rationally tend to believe that the probability that their description is correct grows with the number of witnesses.

More generally, due to the social nature of knowledge, the presence of an authentic agreement between different people (even if non-experts) who belong to the same community strengthens each member's degree of certainty in the relevant belief. On the contrary, lack of consensus leads to a degree of psychologically unwelcome skepticism, which generates doubts as well as a cognitive impasse that must be overcome by a further control.

To repeat this important point, since scientific communities are essentially characterized by the existence of an epistemic consensus about hypotheses based on repeatable observations and indirect control processes (see Chapter 2), intuitively but also rationally we trust science and scientists when they do not disagree.[1] "Intuitively" means that no non-expert is able to provide explicit and complete arguments in favor of the acceptance of a hypothesis that is, however, wholly justified by its theoretical pedigree. With respect to this issue, the historian of science Thomas Kuhn argued that the main difference between a fully developed empirical science (physics is the main example, in this case) and a discipline still remained at a preliminary stage lies in the fact that in the former there is an agreement on the topics that must be dealt with, the methods and solutions that must be implemented to solve the remaining puzzles, etc.[2] On the contrary, in those disciplines and branches of science in which this kind of consensus has not yet been reached—possibly because of the difficulties inherent in the problems involved—disagreement reigns.

An example will illustrate my claim. Among "immature" disciplines, psychoanalysis is often regarded as characterized by a disagreement among at least three different schools: the Freudian, the Jungian, and more recently the cognitivist one. The latter, despite being much closer to the empirical methodology of fully developed sciences, is still hampered by the other two.[3]

Let us suppose that such a controversial judgment on the shaky epistemological status of psychoanalysis were correct. The presence of consensus among the experts of a certain discipline is so important that, on the basis of Kuhn's sociological criterion, one could conclude that psychoanalysis is still at a pre-paradigmatic stage and that, at least from this point of view, cannot be regarded as "fully scientific." Independently of these historical and philosophical considerations, in cases of disagreement between experts and pseudo-experts, citizens are left in doubt, or in a mental state that can easily lead them to a distrust of true experts and of science in general.

Generalizing, given the social nature of knowledge, from a psychological point of view consensus generates epistemological trust in all human beings and hence also in non-specialists toward experts. We have seen why, in the case of science, consensus is a byproduct of an accurate, socially based cross-control of hypotheses. This fact justifies the view that science provides valid knowledge for everybody, hence for non-experts, too. It is exactly on our psychological tendency to overcome uncertainty that the disinformation mechanisms rely, thus creating ad hoc unanimity on false beliefs that conflict with scientifically tested hypotheses.

There is, however, a third, more important reason explaining both the fact that "science works" and the reason why, after a period of constructive skepticism and disagreement, the community of scientists reaches a consensus among its members. This reason, which is philosophically still controversial, consists in the fact that science describes a mind-independent world. Given its capacity to describe the world objectively, science justifies in a much more general way our epistemological trust. We are led to such a pretheoretical belief by the fact that both intuitively and indistinctly we perceive and feel that, out there, there is a world that for the large part is independent of our will and often resists our efforts to use it or improve it for our own advantage. By following Freud, one could even say that what opposes these efforts and subjective wishes is at the origin of the idea that there is a reality independent of our desires, springing from the more instinctual and primitive parts of our brain.

Therefore, also what is referred to as "the person in the street" tends to think that science describes (and ought to describe) a world that is independent of our values, interests, desires, fears, and hopes. In his essay, "The Place of Science in a Liberal Education," Bertrand Russell gave an eloquent philosophical expression to this view:

The scientific attitude of mind involves a sweeping away of all other desires in the interests of the desire to know [... until we become ...] able to see it frankly, without preconceptions, without bias, without any wish except to see it as it is, and without any belief that what it is must be determined by some relation, positive or negative, to what we should like it to be, or to what we can easily imagine it to be.

Russell 1917: 73

The belief that science is successful in carrying out the attempt to describe a mind-independent world—a view that is typical of the so-called scientific realism—is a third way to explain why it is rational to trust the knowledge of experts and scientists. Such a belief, which we tend to subscribe to both psychologically and intuitively, is particularly relevant for my arguments. Even if there is no space here to discuss the complex philosophical arguments supporting scientific realism,[4] for my purposes it is enough to reiterate the fact that the "default" belief that we all have is that the natural sciences that the natural sciences grasp aspects of the world that do not depend on our actual presence on Earth, such as the simple fact that the Sun warms the Earth. Physicists will then explain this by bringing to bear the nuclear fusion of hydrogen taking place at a certain temperature in the nucleus of our star.

Analogously, we believe that there are viruses whose existence is independent of us. To make such statements, we and the scientists need to be equipped with suitable *concepts* (e.g., sun, hydrogen, fusion, virus, etc.), which are *mental* entities. Such a necessary dependence on our concepts, however, does not entail that the entity they refer to is a product or a social construction, but can rather be compared to the need of a telescope—metaphorically speaking, the concepts—to observe lunar craters or the shape of the Moon (the mind-independent physical world). The craters, however, existed on the Moon well before Galileo discovered them with his telescope, as he magisterially illustrated in his *Sidereus Nuncius*, published more than four centuries ago (1610), and shown in Figure 5.1. By travelling to the Moon, we can now directly verify the existence of lunar valleys and mountains.

Going still further back in time, we can reliably claim that the Moon existed before the appearance of homo sapiens on Earth: thanks to samples brought to Earth by astronauts and successive isotopic analyses, we know that the Moon was formed approximately 50 billions years ago. Generalizing, these kinds of

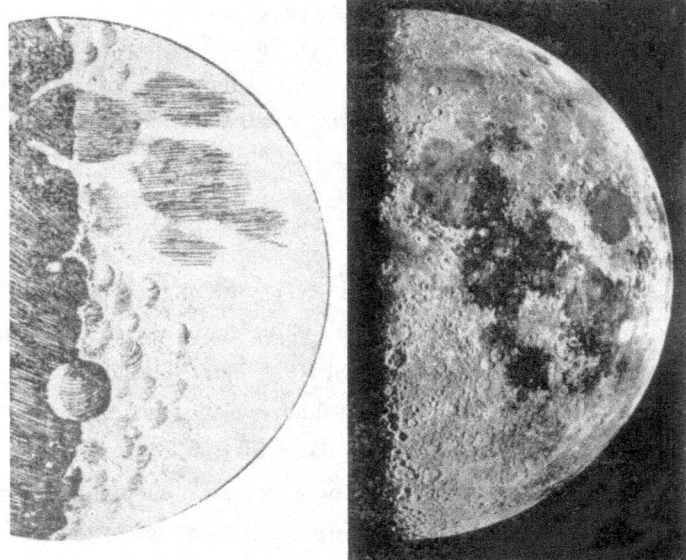

Figure 5.1 The Moon as painted by Galileo in his *Sidereus Nuncius*, published in 1610 (left), compared to a modern photography of the same side. Source: Wikimedia Commons. Online at: https://commons.wikimedia.org/w/index.php?curid=282878 (accessed August 9, 2022).

arguments show that well-confirmed scientific hypotheses correspond, at least approximately, to how things really are, independently of how we would like them to be and, hence, independently of any scientist's individual interests. Such theories must be rationally accepted in the same sense in which we cannot deny the existence of the pull of gravity.

This form of strong objectivity in principle allows us to explain why there is consensus among scientists (my second reason above, a weaker form of objectivity that is, intersubjective validity) without bringing to bear only sociological factors. Scientific beliefs are not social constructs that have nothing to do with the structure of reality, nor are they a mere reflection of power relations and economic interests.

If the acceptance of a theory were not suggested by "stubborn" facts and confirmed hypotheses and we refused to accept the "objectivity" of science in the two senses of the word discussed above, merely partial interests of individuals or groups would prevail. The fact that paracetamol lowers the body temperature and antibiotics are effective with non-resistant infections has nothing to do with the interests of pharmaceutical companies. This would remain true even if the funds provided by the company to do research had been made available just to increase its profit or, equivalently, even if the scientists employed by the company were only motivated by their desire to become famous and rich.[5]

Furthermore, by limiting the power of radically sociological explanations, we can also justify the fact the citizens of a pluralistic democracy, even if motivated by different and often conflicting interests, can share the belief in the existence of a realm of objective social facts. This implies that any public decision must be inspired by the recognition that disagreements about the implementation of a given policy must depend exclusively on the sphere of values: those who share the same political and economic values ought to disagree only about the most effective means to reach their goals.

Psychological or sociological reasons trying to explain the origin of a scientific hypothesis are irrelevant for establishing whether the corresponding theories are well supported by data and therefore can be regarded as (approximatively) true. Consequently, the objectivity of scientific theories as well as our trust in the experts should not depend on the naive belief that the only interest moving scientific research is to ascertain the truth.

5.2 Swindlers and the Origin of Disagreement Among Experts

After having shown the three main reasons why, in areas far from our competencies, we "naturally" tend to trust scientists, it is difficult to explain why disagreement among experts arises, especially in the implementation of policies that have an important social impact, with the resulting distrust of science that characterizes large sectors of public opinion. I shall devote this section to providing some possible explanations for this fact. In the last part of the chapter, I shall delineate some methodological strategies that could allow citizens to understand which group of scientists is to be trusted, thus opposing manipulations due to the circulation of false hypotheses divulged as truly scientific across the social media also by scientists.

There are various reasons (I identified four) that can explain the origin of attempts to instil doubts or ungrounded certainties in the public opinion that end up being perceived as some sort of *explicit disagreement* among experts. The first three are strictly linked, as they are all based on a deliberate activity of scientific disinformation taking advantage of the low level of scientific (and humanistic) knowledge even among the populations of industrialized countries.

The first reason for a perceived disagreement is caused by the intervention in the public debate of swindlers. The second is due to the actual, objective ignorance of the consequences of a yet-to-be-confirmed hypothesis, in which propaganda aimed at the public disinformation on the part of scientists who are not specialists makes its way. By recalling historical episodes already brought to light by other scholars,[6] the third reason will force me to consider another, different aspect of the problem: even when faced with objective evidence in favor of one of the two conflicting theses, it may happen that some authoritative scientists deliberately generate doubts and controversies that delay or veto, for example, some health measures of public relevance. The fourth reason for distrust of experts is more socio-psychological because it arises from a sort of resentment, a psychological need to eliminate any possible distinction between competent and incompetent people. Such hostility towards every sort of epistemic differentiation between the experts (seen as an elite at the service of powerful people) and the common people (understood as an undifferentiated entity that can give its opinion on everything) is evident on many occasions but is often due to an exaggerated prejudice that no scientist is impartial.[7] I shall now discuss these four reasons in turn.

5.3 Swindlers

The first reason giving rise to distrust in experts is caused by the fact that a group of truly competent people is put in contraposition to a group of charlatans who are often sure, in good or bad faith, about the soundness of their views. This is not a case of real disagreement among *experts*, but something that is *perceived* as such because of the lack of widespread scientific knowledge and the oftentimes bad service of some press and social media. Swindlers have a very superficial knowledge of each scientific field they talk about. That is, they did not study the theories and methodologies in the relevant field but are extremely well equipped with a capacity to pick up and use the jargon of true experts, who dedicated a great part of their life to the actual study of the discipline.

Let us imagine a public debate in which a very shy expert and a very confident swindler are put one against the other in front of a TV camera or a webcam as if they were equally competent scientists with opposite opinions. Let us also imagine that each is given the same amount of time, as if they were politicians in an electoral TV confrontation. The habit of inviting people with different opinions involving science independently of their CV is widespread but has no justification, since it makes us forget that the scientists' hypotheses and beliefs can only be evaluated by other scientists and peers, not by incompetents. Scientific hypotheses must be backed up by logical evidence and do not necessarily correspond to common opinions amplified by swindlers.

The deplorable practice of the media to generate pseudo-debates between experts and incompetents on a scientific topic generates the opinion that people who have spent their whole life studying a particular scientific topic and arrogant incompetents are on the same level. After all, it would be as if we endorsed the idea that theft is better than honest toil or, in other words, that studying and working hard to reach important cultural and educative objectives were pointless.

How to react to these frequent events? On the one hand, scientists should not accept the practice of public debates with incompetent people; on the other, there is the professional duty and commitment to defend and argue for the truth: the choice between these two options is not at all easy. Moreover, considering that books and newspapers are read by a constantly decreasing number of people, the eloquence and rhetoric of swindlers can make the difference, so that their view, which contrasts with what science tells us, has more followers than deserved. With a deeper level of scientific information, this threatening kind of disagreement would not be so influential.

Defining swindlers is thus not difficult for scientists and identifying them is rather easy: one only needs to look at the topics on which they provide their own opinions. The actual space given by the media depends on the fact that their provocative views in defense of telepathy, telekinesis, astral influxes, miraculous healing, and the like increases the number of TV watchers. Despite—as well as because of—the lack of any empirical evidence, these views meet our strong fascination for mystery. Defending these pseudosciences, often on the ground that "there is something that goes beyond science, because science cannot understand all phenomena," gratifies our narcissism, as it illusorily strengthens our conviction to be able to control reality and to almost act on it, magically breaking the laws of nature.

A different aspect of the problem is given by superstitious beliefs in magical healing, since they meet the much more understandable hope of becoming healthier again, on which some self-proclaimed experts speculate in a criminal manner. For instance, between 1997 and 1998, the Italian mass media focused on the so-called "Di Bella method," which was based on a therapeutical protocol of Luigi Di Bella in treating cancer that lacked any scientific evidence and that was alternative to those adopted by standard medicine. The same phenomena happened again ten years later, with the so-called "Stamina method," which according to its founder, Davide Vannoni, would have had the power to treat diseases such as ictus, spinal lesions, and degenerative pathologies of the nervous system (e.g., Parkinson's disease and multiple sclerosis). Even in this case, some TV programs promoted the idea that the choice to adopt that medical approach *should be free*: once again, the crucial theme of our autonomy of choice, which in this case was dishonestly promoted by non-scientific theories, came into play. After the official expulsion by the scientific community, Vannoni was prosecuted, and he is now in home detention. As it is clear from these two examples, the scientific community immediately or at least very quickly unmasks swindlers and false prophets, as Kohn calls them (Kohn 1986), but it would be much more advisable to provide citizens—and, more generally, non-specialists—with the tools that would allow them to side sooner with true experts, thus avoiding irreparable damage.

Unfortunately, most of these "magical" beliefs are refutable with only "one unrepeatable irreparable experiment," namely the death of thousands of people. In 1905, the Germans tried to suppress the Maji Maji Rebellion in Tanganyika (modern-day Tanzania). Despite being more numerous than the Germans, the indigenous people were poorly equipped. The shaman who led the revolt insisted on the fact that they could become invincible thanks to a magical potion prepared by him and made up of water and millet. Thousands of indigenous people were killed.

One could object to this claim by recalling what I argued above: the progress of science often depends on disagreement and open discussions of hypotheses that have, at least temporarily, a dubious epistemological status. Why not then consider telekinesis as a possible source for the growth of knowledge? The reply to this objection is that a long apprenticeship is needed before expressing skepticism about scientific theories. This critical activity is possible in two cases: when these theories are still not accepted because they are either dubious or rely on wholly unclear conceptual grounds. As far as the second reason is concerned, Einstein was entitled to criticize, as he did for many years, the conceptual bases of quantum mechanics (despite his recognition of its indubitable empirical success), given that he was one of the creators of the theory. His skepticism let to theoretical and experimental progress (see Chapter, n. 41). It is amusing to notice that even very serious scientific journals receive a few submissions whose main aim is to show that Einstein was wrong on almost everything![8]

A second reason that can explain the disagreement on political decisions is that they *objectively* lack sufficient evidence. In cases of underdetermination of theories by data,[9]—where alternative theoretical hypotheses that all compatible with the same data are insufficient to choose among them—there are various possibilities that must be considered. The first is that scientists can honestly announce to the public that the available evidence is not sufficient to decide among two or more incompatible hypotheses on which the solution of an important social problem depends. Of course, in this case, no dissent is present since the experts agree on their ignorance. Even in the temporary absence of confirmed hypotheses, it could be argued that the history of science supports the claim that eventually only one of the two theories turns out to be correct. Naturally, this is cold comfort in those circumstances in which an urgent decision must be taken: decisions causing less harm should be preferred.

Another, third possibility that might explain the origin of scientific disagreement occurs when the objectively existing underdetermination of theories by data is used to defend the incorrect claim that only *one* of the competing theories (say theory T) is correct. Roughly, in these cases there are two possibilities. If the defenders of T are aware of the underdetermination and are honest, the dissent could be caused by incompetence. If they are dishonest, there will be a conflict between the prudent, honest experts who propends for a suspension of judgment and those who take advantage of a lack of sufficient evidence to propagate ungrounded or still immature hypotheses.

A blatant, third case of dishonesty causing disagreement in the public occurs when there is ample evidence in favor of one of the two competing theories belonging to a field of specialization X, but scientists who may be very competent in *another* field F (and, hence, to some extent, incompetent in X) willingly refuse to listen to their more competent colleagues. Reflecting on the last possibility in particular is useful to warn us that scientists are by no means more honest than the rest of human beings, even if their professional deontology invites them to account for, and "respect," objective facts independently of anyone's interests and values (see Russell's quotation on p. 85 above).

5.4 Doubts and Trust

With a variant to the previous case, a *fourth* explanation of the origin of the disagreement among experts evokes the possibility that a group of scientists, however competent and skilled, could defend willingly an empirically unwarranted or even a false hypothesis. In these cases, the latter group does not exploit an objective ignorance or an underdetermination of theories to propagandize their theory.

Even though scientific evidence would have been objectively sufficient for them to believe in hypothesis h, the dishonest group deliberately concealed or weakened the evidence in favor of h (Oreskes and Conway 2010). The fundamental strategy carried out by these scientists is to nourish doubts on hypothesis h for reasons that are totally independent of its epistemic status and, hence, of impartial though somewhat vague rules of choice such as experimental accuracy, simplicity, scope, and explicative strength of one of the two theories.[10] Here we are dealing with a form of scientific disinformation that is deliberately employed by scientists (Kourany and Carrier 2020).

Much discussed by those who have dealt with the relationship between scientific ethics and democracy was the case of the so-called strategy of the tobacco industry, on which historians and philosophers of science have recently focused thanks to the seminal work by the already quoted Oreskes and Conway (2010). The reason why I briefly go over this case that has already been widely discussed in the literature[11] is not only linked to the fact that analogous strategies were put in place in other important cases (for instance vis-à-vis the safety of vaccines against Covid-19) but also to the fact, for me very important, that an artificially created doubt was used to legitimize the citizens' need to choose autonomously what to do in health-related issues.

An article by Roy Norr published in 1952 in an American popular review, *Reader's Digest*, showed in a very convincing way the statistical correlation between the occurrence of smoking and cancer. To the attack based on this evidence, the tobacco industry reacted by combatting combat "science with science" (Oreskes and Conway 2010: 32). In their research, Oreskes and Conway even found a "know-how" book entitled *Bad Science*, which taught how to argue against solid facts by means of attentively planned, distorted methods of inquiry.[12] With the help of falsely impartial but authoritative scientists who knew that the data they were attacking were correct,[13] the tobacco industry funded committees of scientists with the purpose of casting *doubts* on the correlation above. These artificially created uncertainties were put forward to procrastinate political decisions that could have protected public health much earlier. Conway and Oreskes show that the same strategy has been and is now applied not only to global warming, but to "a laundry list of environmental and health concerns, including asbestos, secondhand smoke, acid rain and the ozone hole" (Oreskes and Conway 2010: 6).

The doubt willingly caused by disinformation and scientific dishonesty was followed by a deliberately misleading attitude, which insisted on the fact that, in the absence of clear evidence, *citizens should take their decisions autonomously*. This is a alluring exaltation of the freedom of choice along with all the very serious risks it brings in such cases. If even scientists openly disagree about the fact that smoking is dangerous, and the evidence is difficult to interpret by non-experts, why quit smoking? We go back here to the conflict between the importance of the principle of the citizens' autonomy and the principle of competence: in certain epistemic circumstances, the former takes precedence in an unjustified way.

Oreskes and Conway quote this passage written by two experts nominated by the tobacco industries, Martin J. Cline and Stanley Prusiner: "We believe any proof developed should be presented fully and objectively to the public and the public should then be allowed *to make its own decisions* based on the evidence" (Oreskes and Conway 2010: 32, my emphasis). If there is no clear evidence, free choices become legitimate.[14]

Despite some differences to be noted below, the analogy with the more recent case of vaccines is quite surprising: deliberately created doubts capitalize on the legitimate citizens' need to take autonomous choices about important issues formulated in a highly specialized and therefore incomprehensible language. These unjustified doubts, generated by deliberate misinformation, elicit reactions of this type: "Why should I quit smoking or have my children be vaccinated if a group of experts or expert-looking scientist disagree about the fact that vaccines are dangerous?" In both cases, the conclusion is, respectively: "I am entitled to *autonomously* decide whether to keep on smoking or get an inoculation."

Evidently, the two cases are different because smoking is pleasant and addictive, while vaccines generate fear, given their extremely improbable, hazardous consequences, of which people can be convinced due to misinformation and disinformation. Moreover, in the case of the tobacco industry, doubts were instilled with the help of *some* corrupt scientists, while doubts about the harmfulness of vaccines are more often generated by the unjustified belief that *all* scientists are moved by economic and personal interests.[15] What matters for my purpose, however, is that in both cases we are in the presence of adulterated evidence. And given the generalization of individual doubts, should we not have recourse to people as the last judge on scientific and political matters, as happens in direct democracies and populistic regimes?

As O'Connor and Weatherall note (2019), such an alteration of data was made easier by the fact that collected data, in this kind of research, rely on statistics and probability. In 1954, the scientific popularizer Darren Huff (1954) showed that one can easily lie with statistics: it is enough to use a statistical sample that is not representative and that is assembled with a methodologically unfair method. A phrase that often circulated in the nineteenth century had it that, "There are three types of lies: lies, damned lies and statistics."[16] Just to make an example, to carry out, say, a political poll for a national vote, one should not collect the answers provided only by old citizens with high income and living in the richest city areas. This example illustrates how one can use unreliable forecast methods. We shall see in the next chapter why the defense of the citizen's decisional autonomy necessarily goes through some teachings, even basic ones, of probability and statistics, which can be easily imparted in secondary schools and that highlight the rigorous and wide-ranging applicability of both subjects.

The awareness of the methodological relation existing between causality and probability can be illustrated by the fact that the already quoted Cline, as a witness in the trials against the tobacco industries, made a deceptive remark of the kind: "*we are not certain that tobacco causes cancer.*" The truth of this statement, however, is accepted by anyone familiar with the epistemology of science and probability but should not be interpreted as Cline wanted us to read it.

First, as argued for in Chapter 1, scientific theories are not infallible: we can never be absolutely certain of their truth, that is, they may be true but we cannot be certain about it. Second, it is more important to note a less familiar fact: we can never say with certainty that smoking causes cancer when we consider a *single individual*. Such statements, especially in biomedical sciences, are never possible because of the high variability of individual persons. The notion of probability that enters into the interpretation of the italicized statement involves the *frequency* of a certain attribute in a given class and not, strictly speaking, individual persons. For instance, the probability to die for 71-year-old person is the *ratio* between the group of people who die at that age and the whole group of 71-year-old people. Strictly speaking, then, the probability of death of a *single member* of the class is not defined because a single 71-year-old person may be very healthy and another very sick. It is only for mass phenomena that the probability as frequency is defined, and insurance companies prosper only on *great numbers*.

Consequently, the probability that the above italicized proposition by Cline is true increases proportionally to the number of alternative possible causes that have been eliminated and that could explain the phenomena in question in different ways. For instance, by assembling data on people who smoke many cigarettes a day, live in a polluted area, do not practice any sport, have a certain genetic predisposition for cancer, etc., one discovers that among these people there are more individuals suffering from cancer than among people who live in an equally polluted place, do not practice any sport, have similar genetical lineage but smoke much less or not at all. In other words, smoking *increases the probability of cancer but does not cause it*.

Summing up, the case of the tobacco war clearly illustrates the abovementioned fourth reason giving origin to disagreement among experts and generating perplexity in the public. Such a behavior on the part of scientists is the most unfair in terms of scientific ethics that, as Weber (1949) and Merton (1973) argued, should be based on a neutrality with respect to the personal, political, economic, and social interests (value neutrality or, in Weber's word *Wertfreiheit*, literally "freedom from values"). The fact that these norms are consciously violated implies that scientists are aware of them.

In this context, I cannot deal in its full complexity with the vexed separation between science and values, which pragmatist philosophers typically deny. However, we should keep in mind the following important distinction: one thing is to admit that the choice of a particular field of inquiries by a single scientist or a community of scientists depends on, and is always motivated by, individual or social *interests and values*; quite another is to defend theories that are not supported by facts, this being a behavior that violates what I would refer to as "true neutrality." There are of course values also in science. The case study presented above illustrates in a very clear way the sense in which the application of *epistemic values* like the already mentioned simplicity, consistency between hypotheses and data, experimental accuracy, and the like must be distinguished from non-epistemic values that depend on economic or political interests. The strategy of the tobacco industry is based on an objective distinction between reliable and adulterated evidence and the former is reliable because it is independent on economic or political values, which I will refer to as non-epistemic values. It follows that the best way to defend the neutrality of science consists in highlighting precisely those cases in which it was gravely violated by those scientists who harmed other human beings. Sadly, the history of the strategy of the tobacco industry shows how, only after a few years, the dishonest members of the scientific community also embraced the correct theory. Nonetheless, before freedom from non-epistemic values—the hallmark of science—was reached, much damage had already been done.

A pious preach for the ethics of neutrality is useless, but the distinction between the two kinds of values signaled above is important to fight against false beliefs which are still widespread also thanks to economic interests that cause scientific disinformation. The thinning of the ozone layer is *not* due to volcanic activity but becoming aware of this lie presupposes a radical change in our education system that should reinforce scientific information. Otherwise, how can a collective solution to global warming be provided if until some time ago even some very authoritative scientists moved by non-epistemic interests denied the hypothesis, supported by a large volume of data, that it has an anthropic origin?[17] It would suffice to become aware that if global warming were caused by astronomical factors then it would be much slower than it is. The reduction and elimination in the production of harmful substances into the atmosphere should be encouraged even if (which is *not* the case) we had some epistemic doubts about some of the causes of the tragic phenomena. Admittedly, here the fundamental value of preserving human life suggests a unique course of action, since in such cases waiting for an agreement among scientists (even if doubts were justified, a claim that, in the problem at hand is not to be granted for the problem at hand), would cause the extinction of all forms of life on Earth.

5.5 Scientific Fraud and Inductive Risk: Two Additional Sources of Disagreement

The problem of scientific frauds which is more widespread in the biomedical sciences, might constitute an additional cause of the citizens' perception that science and therefore the world of experts are imbued with doubts and unresolved controversies (Chevassus-au-Louis 2019). What kind of impact does scientific fraud have on the manipulation of public opinion? A strong motivation for scientific fraud or the alteration of scientific data arises from the necessity to obtain funding. To this important element, we must add the necessity to publish in a merely quantitative way, to get positions in universities or research centers. This phenomenon has been generally referred to as "publish or perish."

Enrico Bucci (2015), a former researcher from CNR (Italian National Committee of Research) who has later devoted his carrier to studying scientific fraud, showed the manipulation of images in the biological sciences is easy to make and efficacious in its results. The findings collected by Bucci by means of anonymous questionnaires (which however should be further checked) are striking. In a casual sample of papers on the foundation research on cancer, it is shown that 1 out of 4 contain false or altered data, that 81 percent of biomedical sciences students at the University of San Diego have declared to be ready to use false data to get funding, and that 1 out of 50 students at that university had already partially manipulated scientific data.

Luckily, the growing number of such cases are producing effective immune reactions on the part of intellectually honest scientists. For example, some software tools have been implemented that can highlight the altered images so that scientific frauds are, sooner rather than later, detected. Otherwise, the data reported and classified as fraud would not have been available! This remark can be further justified by some conceptual analysis. The term "fraud" presupposes the existence of an honest way to publish the results of scientific inquiries and that there *is* an objective difference between true or probable and false or improbable hypotheses and conjectures. However, the proposers of a relativist and non-objectivist conception of science deny this difference, paradoxically by relying on the cases of fraud!

On the one hand, one could object that scientific frauds must be classified as a kind of disagreement that affects *only* the scientific communities and that, as such, it is not transmitted to the wider public. Thanks to the social requirement of the experiments' repeatability discussed in the first chapter, experts usually bring to light the falsification of the data.

On the other hand, however, the mere existence of scientific frauds is functional to my main claim. One cannot exclude *a priori* that frauds can be amplified by the social media and can thus reach the citizens, with the result of producing what is perceived as a conflict between "experts" that generates further doubts in the wider public and that justifies the procrastination of political decisions that have an important social impact. The mechanisms of the social establishment of scientific hypotheses discussed in the second chapter, however, take care of the tendency to construe carriers on scientific frauds.

There is a final aspect that needs to be discussed and that concerns the necessity to find a delicate balance between two opposite values that, as we will see, are necessarily involved in the choice between different hypotheses in applied science. The key issue raised by the philosophers who cast doubt on the distinction between facts and values (in the sense already explained) is that a commitment to non-scientific values determines, to some extent, the *result* of scientific experiments. These cases deserve separate treatment since they typically involve cost-benefit analyses or, much more importantly, serious conflicts between economic interests and our health (see, among others, Douglas 2000, 2009; Dorato 2004; Barrotta 2017).

In order to defend the thesis of the inseparability of science and values, it is argued that the epistemic norms that are relevant to evaluate scientific hypotheses (the several times mentioned scope, explanatory power, consistency, etc.) are inseparable from non-epistemic values, such as people's health, or economic and political interests. To assess these kinds of arguments, we must consider the fact that all statistical tests entail an inductive risk, a fact implying, on the one hand, the possibility of taking to be true a false hypothesis and, on the other, taking to be false a true hypothesis (see Hempel 1965: 93). In other words, in both cases, such a problem implies the probability that the hypothesis we are supporting is wrong.

The main idea behind these important arguments is that the kind of test that is used depends on a previous commitment to non-epistemic values. Before discussing Douglas' seminal argument, we must introduce a couple of notions. Suppose that a person suffers from some symptoms and undergoes a test. To reduce the risk of false negatives, we must prepare first a highly sensitive test that "reacts as much as possible" to the pathogen by lowering as much as possible the number of sick people who are regarded as healthy because they were falsely tested negative (false negatives). This means that in very sensitive tests that turn out as positive, the positivity is very probable. After this test, usually, positively tested people take a very *specific* test to make sure that they are really sick. A test is specific if it is made to react as precisely as possible with healthy people. In this case, the purpose of the test is therefore to minimize the number of false positives. It may well happen, however, that this latter objective is unreachable without increasing at the same time the number of false negatives, that is, the number of sick people who are tested negative.[18]

We thus have the following problem: if there are no tests allowing us to make the number of false positives *and* that of false negatives decrease, what should we do? In some cases, the choice is easy: if we do not want to risk transfusions with infected blood, we must trash away sacks of blood that *might be non-infected*, thus opting for (highly sensitive) tests that decrease as much as possible the number of false negatives, and therefore lower the probability that infected blood that is tested negative is transfused.

In cases in which from an epistemic point of view the evidence is not clear, and a balance between two conflicting values cannot be found, the choice between the kind of tests that must be used—and hence about the threshold or cut-off value separating illness from its absence—may indeed depend on a choice based on *non-epistemic* values. Such a choice, which to a certain extent determines the results of the tests, is obviously a potential source of disagreement. For example, when we must assess the harmfulness of a chemical product (Douglas, 2009, gives the example of dioxin), on the one hand, there are the economic interests (clearly non-epistemic) that are caused by the suspension of the industrial production, unemployment, etc. On the other, there are the consequences of keeping on producing the chemical on our health (an obviously more important value that however is also non-epistemic), which makes us tend toward very sensitive tests: the number of sick people who are tested negative is as low as possible.

Given that the epistemic value of these tests (*evidence*) depends on the *choice* between two conflicting non-epistemic values (in this example, health and economic growth), Douglas (2009)—and many philosophers after her—defended her seminal claim that science and non-epistemic values are *inseparable*. If in the assessment of the harmfulness of a product our health is at stake, such a choice should be based on very sensitive tests, such that the probability that a sick person is tested healthy (a case of false negative) is as low as possible. On the other hand, lacking reliable evidence, the defense of economic interests (including low rates of unemployment) will suggest less sensible diagnostic tests in which the disease will appear less dangerous than it is because a greater number of sick people is tested negative.

Douglas argues that in similar cases, the choice between the two methods of testing (increasing sensitivity or specificity) depends in turn on a choice between two non-epistemic values. It is this choice that determines *the probability that a sick person is tested negative or vice versa*: the sentence is italicized because it reports a description of a *fact* (the object of science), that, however, depends on the choice of values (health and unemployment). Whether a test is prepared in such a way that it is more or less sensible depends on our values.

In situations that involve public decisions, if the experts are motivated by opposite values that are not made explicit but that determine the results of the tests, it will be very difficult for the citizens to *decide autonomously*. How can they choose in the absence of solid statistical evidence and without being aware of the conflicting values which the different groups of experts are committed to? One solution, desirable but illusory and idealistic, is that the scientists openly reveal their allegiance to these values (Kitcher seems to defend this claim, 2011), but who would want to deny that one's prior aim is to protect our health? The case of the tobacco industry provides an obvious answer to this rhetorical question: a dishonest scientist will never say that she wants to protect economic interests!

It must be noted that the inductive risk discussed by Douglas is motivated by what I referred to as the second reason that generates disagreement among scientists, namely a *condition of genuine ignorance*. In our case, it is ignorance about the toxic effects of a substance at a *certain moment in time*, caused by the impossibility of lowering the number of false negatives without raising at the same time the number of false positives.

A tragic situation occurs when decisions must be taken urgently. If strong evidence is missing and people's health is implicated, the principle of precaution ought to prevail by deciding to preserve human life. On many occasions, the choice of the "right group of experts" is only apparently easy: how can we be sure that scientists who underestimate the risk of raising the number of false negatives are not moved by personal interest?

The plausibility of my evaluation of inductive risk as presented above can be further motivated by the fact that, in different areas of inquiry, there is very reliable evidence to conclude in favor of the harmfulness of a product or of the actual presence of an illness. In these cases, our knowledge is empirically solid, and therefore a positive result to a test confirms beyond every reasonable doubt the presence of the illness (think about AIDS). In the case of epilepsy, the number of false positives is extremely low: while a negative EEG does not prove the absence of the disease (false negatives), the anomaly of the track is strong evidence for the presence of the disease (no or few false positives).

In any case, Douglas's position is not very distant from mine: she is not a radical sociologist of science, since she admits that, "Hempel is right in holding that whether some piece of evidence is confirmatory of a hypothesis ... is a relationship in which value judgments have no role" (Douglas 2000: 564). To me, this is the fundamental point: although in certain circumstances conflicting non-epistemic values do exist, the relationship between evidence and hypotheses is objective (despite the opinions of groups of self-alleged experts). Moreover, the calculation of the consequences of a given decision presupposes the epistemic value of objectivity since it is thanks to experimental results that the principle of precaution can be invoked. Likewise, the fact that in certain circumstances, decreasing the number of false positives and negatives at the same time is impossible, is also an objective fact.

Finally, there are conflicts between the political authority and the individual. The web and social networks, while letting narcissism thrive thanks to the fact that everyone can express their opinion on everything (science, literary criticism, history, etc.), quickly propagate false theories spreading as aggressive viruses that favor the partial or total elimination of any filter between experts, scientists, and web users. Consequently, lacking scientific literacy, the opinion of the former loses any authority. This apparent equality in the distribution of epistemic resources gives the citizen another motivation to want to directly choose the scientific policy that appears to favor her autonomy of choice. Why should she be obliged on the part of governmental authority if the latter relies on experts whose opinion is as warranted as any other one?

This can also explain why, partly motivated by an opposition to state authority, some parents want to "autonomously" decide not to vaccinate their children against measles or Covid-19, as if this were a legitimate expression of their freedom of choice. Of course, these are not decisions that concern only individual citizens that, as such, would be fully legitimate. In the case of measles and Covid-19, an autonomous decision based on beliefs that are devoid of any scientific evidence has relevant social consequences, since it exposes to a serious risk not only the denialist's life but also that of other members of the society. Before taking any autonomous decision that could put the health of others in danger, one should calculate the inductive risk that I mentioned above, that is, that the theory one supports, which is taken as true, is in fact false, and conversely. To know something about the inductive risk would be important for us all!

In sum, if we put these problems into a wider perspective, the distrust for experts, joined by the more than justifiable aspiration to make choices that are as autonomous as possible (recall Kant's message) has two potential consequences, which are dangerous both for the citizens' health and for the survival of democracy. Both are characterized by the risk of encouraging frequent recourses to the opinions of citizens about scientific and political matters and, hence, to radical forms of direct democracy openly aiming to exalt the citizen's autonomy beyond measure. I can now support in more detail a thesis that I anticipated in Chapter 1: in direct democracies, the subversion of the principle of representation embodied by the mediation of experts reduces the principle of autonomy to dichotomic answers (yes or no) to very complex questions that instead require a competence that can be acquired only after many years of study. Analogously, political, not-ethical decisions taken through a referendum require a yes or no to issues that are not raised by us but by those who are in charge. We have seen why in social epistemology the problem of choosing the "right experts" is of crucial importance (Goldman 2001): how can we *trust* experts and scientists if we do not know whether their disagreement is due to merely personal interests?

What I suggested in this chapter is that a clarification of the nature of the problem requires first an inquiry into its different causes. In the next chapter, we shall see why, as anticipated by Dewey more than a century ago,[19] its solution presupposes a global reform in our education systems aimed at developing the problem-solving attitude toward social and ethical problems that is best expressed by the empirical method of science.

6

How to Navigate in the Disagreement of Experts

The Need for Greater Scientific Literacy

In this chapter and the next, I will show why an increase in scientific literacy is a necessary condition for citizens and the public opinion to be able to decide autonomously and rationally—thereby avoiding manipulation—about social policies involving scientific theories, which is thesis T_2 announced in the Introduction. Indeed, it will become even clearer why, in societies characterized by increasing technological developments, without a deeper awareness of the empirical dimensions of social problems, a democracy worth its name cannot exist. As we shall see, progress in scientific literacy, accompanied by opportune and important openings to the humanistic culture, is also the best way to take sides on issues involving a dissent between true and alleged experts.

More schematically, I will articulate the defense of T_2 in the following way. In section 6.1, I will show why Condorcet's jury theorem, which can be dealt with very simply, proves illuminating for the issues I am concerned with, as it rigorously highlights the consequences of disinformation for the functioning of democracies *merely* based on the principle of majority. By reviewing some of the most discussed claims in the current literature, in 6.2 I will offer a more precise characterization of the meaning of "scientific literacy" by raising the key question: what should citizens be scientifically literate *about*? In 6.3, I will highlight some very worrisome data on the educational level of European and North American citizens.[1] This fact is certainly due to situations of social hardship of many economically depressed geographical regions that must be fought through political decisions of greater investments in education.

Greater scientific literacy would also strengthen the belief that the choice of representative forms of democracy is preferable and inevitable, which is thesis

T₁ announced in the Introduction. The first factor justifying this statement is given by the facts that, (i) greater familiarity with scientific knowledge brings an awareness of its specialized nature and therefore, (ii) of the importance of the principle of competence. The latter factor depends on the former. The fact that a research *community* A believes and trusts the theories adopted by a community B speaking a different language depends on the fact that both communities base their hypotheses on the *same* kind of control of empirical hypotheses. Here "same" does not necessarily involve a "fetishism" about *the* methodology of science,[2] but rather an attention to the empirical dimension of knowledge as it is constantly checked by the socially based epistemic rules highlighted in Chapter 2. Consequently, as seen above, the reliability of the citizen's beliefs in the effectiveness of certain policies implies a principle of delegation to the scientific knowledge developed by true experts that is neither blind nor irrational. These epistemological factors decisively justify the acceptance of the only forms of democracy being compatible with them, that is, *the representative ones* based on the principle of delegation.

Finally, in section 6.4, I will suggest a simple strategy for assessing the scientific authority of the protagonists of a public debate on scientific issues, based on information that is not entirely unreliable and quickly acquired through the Internet. As we will see, the elementary nature of this strategy is not incompatible with its effectiveness, but it is only a first step toward the solution of a very complicated problem.

6.1 The Importance of Scientific Literacy for Democracy: Condorcet's Jury Theorem

The proper functioning of a democratic society pervaded by hyper-specialization, regulated by the majority principle, and at the same time respectful of the pluralism of values, is inextricably linked to the probability that our opinions are close to being correct. It is *only* in this case that the majority principle can provide us with reliable results. This fact was implicitly argued for in a rigorous way more than two centuries ago by the French mathematician and philosopher Nicolas de Condorcet (1743–94),[3] one of the first who tried to apply mathematical models to social problems.

Here I will reformulate Condorcet's theorem by adapting it for our purposes but without distorting its meaning and logic. In this essay, Condorcet asks us to imagine jurors who must decide whether a defendant is guilty or innocent (Condorcet 1785). From our perspective, I can translate this problem into the one posed at the beginning of the present chapter. The aim is to use Condorcet's result to understand which cognitive conditions are indispensable to enable the jurors (out of the metaphor, the citizens) to choose as independently as possible between two opposing theses (i.e., scientific hypotheses defended by two different research groups or experts), one defended by the defense attorney and the other by the prosecutor (we can imagine her being a dishonest or self-styled specialist). It is important to keep in mind that in my use of the theorem, the defendant stands for the group of the "true" or reliable experts, while the prosecutor represents the group of "pseudo-experts."

Suppose that the jurors belonging to group A are better-informed and that those in group B are less informed. It does not matter whether the defense attorney's hypothesis supporting the innocence is right or wrong but suppose that the defense attorney is right and that she is a true expert. What we want to show is that group A will eventually side with her judgment because its members are better-informed and will identify very reliable criteria for their conclusion: the accused is ("very probably") innocent.[4] We will also show that group B, composed of less-knowledgeable members, will side with the wrong thesis ("very probably") guilty, which is defended by the incompetent or uninformed prosecutor, and therefore is not backed up by reliable evidence.[5] In other words, they will choose the "false experts." The problem that the jurors must solve is to choose *independently* between the two verdicts, siding in favor of innocence or guilt individually, that is by evaluating as autonomously as possible the arguments of the defense and of the prosecution, respectively.

The premises of Condorcet's jury theorem are the following:

(a) only one of the two options is correct, that is, either the belief of the expert members of group A or those defended by the members of group B;

(b) the majority of one of the two groups of jurors/citizens (A in our case) is more than 50% likely to have the correct belief according to them, the probability p of the truth of their belief is $p(A) \geq \frac{1}{2}$. This premise expresses in a precise way the *principle of competence*;

(c) no opinion of a juror influences that of the other, which means that the probability of the correctness of the opinion of one juror is *independent* of that of the other: each juror forms her judgment independently of the others;[6] this premise corresponds to the *principle of autonomy of choice*.

Before stating the conclusion, let us discuss the truth of the premises. The first and second premises are plausible, whereas the third is more controversial and therefore deserves particular attention. The first premise states that the accused is either innocent or guilty (theory A is true or false), and the assumptions leading to the verdict are either true or false (the principle of the excluded middle holds).[7]

Premise (b) states that before the trial or the public debate, the jurors (citizens) do not know with certainty which of the two hypotheses (the beliefs of the research communities or the group of experts) is correct. However, they have opinions about their correctness that is expressed by different "degrees of probability" of being true.

It is important to specify that the probabilities p mentioned in premise (b) are different from the probabilities of getting tails in the toss of a coin, which is defined as the number of favorable cases divided by the number of possible cases. Nor can these probabilities be regarded (for example) as the *frequency* of the number of born male or female in a very large sample of newborns, an approach to probability that was briefly discussed in the previous chapter concerning the conceptual link between causality and probability of cancers for smokers. The probability that intervenes in the jury theorem is defined as the "credence" or degree of belief with which an individual "bets" on a given hypothesis.[8] This degree of belief depends on the reliability of the juror's background knowledge and is influenced by the pleadings of the defense and the prosecution, so that the jurors can make up their minds on the basis of the evidence brought by the prosecution and the defense.

The third premise (c) entails that each juror (citizen or individual scientists) forms her opinion *independently* of the other members; in the case of a jury, the opinion of each juror remains unchanged after a dialogue with any other juror, that is, each forms her opinion independently of the others.

By assuming the truth of these three premises, it can then be shown that:

1. If the beliefs of the majority of the group A of jurors are more likely to be correct than wrong (that is, their strenght is above 50%),[9] and each forms her belief independently of the others, the probability of voting for the correct verdict increases with the number of voters, until it reaches certainty as their number grows indefinitely. In other words, on the basis of the three premises above, the final verdict "innocent" voted by the better-informed jurors/citizens is correct and the "true expert" (the defendant) is identified. In this case, the majority's opinion is correct and the principle of choosing by the majority rule is justified.

Keeping the premises, (a) and (c) unchanged but assuming (b') (i.e. the probability that citizens' opinions in group B are correct is *less than 50%*), it follows that most citizens are more likely to be wrong and, as the number of voters grows, the likelihood that their opinions is correct (so that the citizens embrace the theory defended by true experts) becomes lower and lower and tends to zero.

Paradoxically, on hypothesis (b'), unlike what happens with (b), the opinion of a randomly chosen individual belonging to the original group B is more likely to be correct than the opinion of the majority made up of *very many* individuals, given that the probability that the opinion of the majority is correct tends to zero with the increase of the number of voters!

The theorem tells us that if jurors and citizens are on average competent (assumption translated in premise (b)), and do not consult each other in a way that prevents their beliefs from influencing each other (c), then the final decision of the jury (the community of citizens) approaches the truth as the number of jurors grows indefinitely.

Consequently, better-informed citizens (group A) are more capable of identifying the group composed of real experts and therefore justifiably trust their hypothesis. They are far more likely to make the true decision. With the limits that we will now discuss, *the theorem demonstrates that a democracy based on the principle of majority*, which expresses the will of the people, *works only on the condition that the average citizen is competent and forms her opinion independently of those of others*. On the contrary, an individual chosen at random who is part of the group B (in which, according to premise (b'), the opinions are more likely to be false than true) has more correct opinions than those of a very large majority. In fact, with the increase of the number of members of B, the probability that one of its members chooses the right hypothesis and the group of true experts supporting it is null. In the original case of Condorcet's jury theorem, the verdict on the accused will be wrong: if the person is innocent, she will be convicted, if the person is guilty, she will be acquitted.

However, there are at least *three* objections that can be made to this theorem and therefore to its use in the present context. The first is that its results are trivial: one does not need a formal argument of this sort to prove such an obvious conclusion. This objection, however, can easily be refuted. If even in the history of mathematics, apparently intuitive hypotheses have turned out to be false, we must admit that science, in general, is characterized by a process of emancipation from intuitions and common sense, which in crucial cases are refuted. Demonstrating in a precise way, albeit through an abstract model, intuitions of which we are sufficiently convinced still constitutes a form of cognitive progress. Moreover, the conclusion that an opinion randomly taken by the uninformed group of people is more likely to be correct than the one taken by an extremely large uninformed group would not have been easily reached without the theorem. The same remark holds for the convergence to certainty as a function of the number of people possessing correct information.

The second, more serious objection, consists in the fact that the premises of the theorem are not realistic. The unrealistic aspect revealed by this second objection would be due to citizens or jurors being principally unable to have all the skills and knowledge necessary to assign any degree of probability to hypotheses put forth by the defendant of the prosecutor (that is by the group of true experts A against the self-alleged ones B): we know that knowledge is not equally distributed. The most correct way to assign probabilities to the jurors' opinions would then be to consider them all exactly to be equal to ½ (mere chance): in the case of absolute ignorance, no hypothesis can be considered more probable than the other and in order to make up one's mind, one could as well toss a fair coin. Consequently, the second premise of the theorem (that the juror's belief of group A is greater than ½) could not be endorsed since it would be false.

However, this issue has already been partially addressed in the previous chapters, particularly in Chapter 2, where we saw that all our beliefs are "nodes" belonging to a very complex network. Checking the correctness of our beliefs can only be *indirect*. If the jurors'/ citizens' problem is to decide to which group to delegate the task of giving correct answers to social problems (whom we should trust), the need for a higher scientific literacy and cultural level in general emerges even more clearly. It is only on these conditions that the probabilities they assign to the various hypotheses are closer to the truth. Once again, this aim can only be reached by a major reform of our educational systems that can train our mind to solve social problems by using the trial-and-error method of science.

The third, much more serious problem is that, against premise (c), before reaching a decision, not only do the jurors actually influence each other when the sessions are over, but they *ought to do so*. As we have seen, if we put the citizens of a deliberative democracy based on public discussion in the place of the jurors, this element becomes essential. In this regard, premise (c) of the theorem has been regarded as unacceptable: "a free press, a public discussion and therefore a mutual influence before the vote are not accidental characteristics of a democracy. Without access to information and opinion sharing sources that go beyond their immediate knowledge, voters would be uninformed and lacking in reference points" (Anderson 2006: 11).

Anderson's judgment is initially plausible, but there are ways of defending premise (c) in a question that he does not consider. For example, it has been observed that before the vote—that is, in our reinterpretation, before choosing which group of experts to rely on—citizens ought to consult and inform themselves as much as they want. It is only at the time of voting (that is, when they have already decided) that they must no longer negotiate with others and keep their own opinion.[0]

In other words, the answer to the objection to (c) consists in noting that the probabilities at stake at the time of the vote (and therefore of the final choice of the defense or prosecution thesis) are the result of a long, previous debate among the jurors/citizens. There would thus be *two* phases in the process of forming an opinion, one preceding the other. In the former, there is a mutual influence between the beliefs of the jurors/citizens as it must be: in this way, they can learn from each other and are connected to information sources. In the latter phase, the beliefs thus formed (or rather the strength/probability with which they are believed to be true) reach a certain stability, become independent of the viewpoints of the other jurors, and no longer change according to those of others. This feature attempts to embody, as far as possible, the principle of intellectual autonomy, fundamental in every democracy and which I mentioned before when referring to Kant. The condition of independence represented by (c) would therefore reflect the decision not to change one's opinion at a certain point in the investigation. This does not imply that these beliefs were not formed through a process of public dialogue and mutual influence, just as it ought to be in a democracy and within a scientific community.

What is central in the justification of the first stage implied premise (c) is the principle of dissent in a community of scientists or citizens, on which I insisted in Chapter 2, since it is this principle that increases the chance to arrive at the correct theory. In Keren's words: "norms against following the consensus have a unique function in promoting reliability. For scientists who simply follow the majority view no longer contribute to its reliability" (2018: 787). The independence of opinions sanctioned by (c) is a necessary condition for the existence of dissent.

In any case, regardless of the difficulties in extending the validity of the third premise to cases that are much more complex than those concerning a jury, arguments of this type highlight the importance of possessing reliable information to decide which of two (or more) groups of experts is plausibly more informed (premise b of the theorem). Furthermore, they shed light on the necessary conditions for the majority principle to be cognitively grounded, given that the probability of having a correct judgment increases with the number of voters/citizens.

Interestingly, in these situations, the answer to the question, "Why should I vote?" if asked by well-informed jurors, is that by doing so, voters who have a sufficient competence would contribute, even if by little, to the likelihood that the final verdict is correct, since this depends on the number of voters. Finally, the considerations already made regarding the tyranny of the majority are confirmed by the fact that, as the number of voters increases, an ignorant majority in the long run will almost certainly decide against the well-being of *all* the members of the society. This is a consequence of the fact that the relevant decision is based on erroneous beliefs. Since beliefs are the main guide to the success of our actions, it is indispensable to favor a greater literacy of the population and in any case to let a minority that could be right have its say.

6.2 What Should Citizens be Scientifically Literate About?[11]

I should consider first some objection to the claim—to which I gave so much emphasis—that greater scientific information does not amount just to a rhetoric refrain or an empty political slogan (Feinstein 2010). Even a brief review of a very small part of the immense literature on scientific literacy[12] will be helpful to compare my position with some of the most discussed theses to be found in current debates. I will start with the useful three-partite distinction of "scientific literacy" proposed by Shen (1975): *Practical, Civic,* and *Cultural* literacy. The first involves applied or technical knowledge that, according to Shen, is useful "because it can immediately be put to use to improve our living standards" (ibid.: 266).[13] This form of literacy is less important for my aim, even though it is important to teach the older generation how to use new and beneficial technological devices.

The second is instead of central significance, since it is supposed to "enable the citizens to participate more fully in the democratic processes of an increasingly technological society" (ibid.: 266). The third, which involves the place of science in a unified view of culture, is only apparently less important to achieve the second aim, and will be discussed in the next section.

Feinstein, for one, argues that the usefulness of scientific literacy ought to involve the use of science in ordinary life, in terms of the so-called *public engagement with science*. His interest is therefore in the *second* form of literacy. He proposes a rapprochement between scientists and laypeople by stressing that the latter should become "*competent outsiders*," a notion entailing the possibility to discover which aspects of public decision-making is more relevant for the citizens (Feinstein 2010: 180; Keren 2018). The name "outsider" stands for the fact that the average citizen cannot become expert in any specialized field, but she can nevertheless become competent enough by learning how and why some scientific concepts are relevant to solve their daily life problems, social problems included. It must be admitted that the distinction proposed by Feinstein between the claim "knowing that science is *good*" and the claim that "knowing that science is useful" is correct (Kitcher 2001; Keren 2018) and that the latter aspect is more relevant to decide between two different policies.

However, the number of topics that "civically literate" outsiders should be competent about is very large: genetically modified organisms, nuclear plants, embryonic stem cells, remedies for pandemics, etc. It seems to me plausible to assume that without an awareness of how a scientific theory in general is confirmed, it will be difficult to make up one's mind in matters of great social interest (vaccines, anthropogenic climate change, etc.). This awareness can be achieved without becoming "insiders experts" in some of these fields.

Even a passing acquaintance with statistics, for example—on whose importance I will focus on in the next chapter—is a case in which the *methodology of science* can play an essential role, since statistics is present in all empirical sciences. We should then raise the question which methodological rule can be regarded at least as partially necessary for the citizen to become a "competent outsider" in Feinstein's and Keren's sense. For instance, if we are told that only in one in several thousand times can an injection cause lethal effects, the decision about whether to undergo a treatment that has been successful in a great number of cases should be straightforward. The same remarks hold for the illusion that geologists should be able to predict *exactly when* the next earthquake is going to occur, which can be dispelled by some knowledge of probability theory. Each earthquake is different from the others and there is no probability of single events!

Huxster et. al.'s (2018) insistence on the epistemic notion of *understanding* and its difference from the stronger notion of *knowledge* is an extremely important attempt to clarify what scientific literacy means, what its aims are and in what sense it relates to Shen's "Civic Scientific Literacy." As Huxster et al. themselves note, the task of understanding what "understanding" means in "public understanding of science" is extremely complex and should be the object of future studies. Their approach is essentially holistic: one understands a subject when one is capable of drawing inferences and connections with other cognate fields (ibid.: 759); it is the fact that "understanding involves the ability to *work with* information that makes it so well suited to capturing the concept of scientific literacy" (ibid.: 767, emphasis in the original). Since *understanding* is a weaker epistemic requirement than *knowing*, it seems plausible to claim that citizens can achieve the former epistemic state and become competent outsiders without being able to achieve the latter state. The pragmatic component of understanding so defined by Huxster et al. would enable citizens to actively intervene in public debates.

There are a couple of points that I want to raise as possible objections to this remarkable paper: (1) even though it might be true that one can *understand* a subject S without *knowing* the main facts about S, why would knowledge not be so holistic as to prevent someone from doing something with what one understands? (2) given that the notion of "understanding" is relevant not only in the context of public science but also for the epistemology of science, why does their approach to "*public* understanding of science" not take into account the contributions of philosophers of science to clarify the notion of "*Scientific* Understanding of Science" (de Regt 2017). The latter is a general remark about the whole literature in the field.

6.3 Functional Illiteracy, Relapse into Illiteracy, and Cultural Scientific Literacy

Suggestions on *how* to achieve a widespread scientific literacy here are not appropriate but proposing some food for thought is important to realize that the goal is achievable. In order to increase the number of students enrolled in scientific and engineering faculties—an obviously necessary step—historians and philosophers of science can play an important role. Shedding light on the foundations and the development of the natural sciences will provide a deeper understanding of their subject matter. In addition, an increased presence of historians and philosophers of science in scientific curricula would positively affect future pure research and technological spinoffs[14] and, more importantly, also the quantity and quality of teachers in high schools, the number of good science communicators, etc.

Despite these plausible considerations, that I will develop in the next, final chapter, there are two factors that in the present context I omitted to consider:

6.3.1 The relevance of *ethical and religious* considerations in attitudes toward science even among people that are well informed about it
6.3.2 The presence of functional illiteracy also in developed countries.

Let me discuss them in turn.

6.3.1 The Relevance of *Ethical and Religious* Considerations in Attitudes Toward Science even Among People Well Informed About It[15]

Empirical studies have shown that scientific literacy can be *positively correlated* with phenomena of polarization: this occurs when ideological and religious commitments intervene *because* of more scientific information! "political and religious polarization over science and technology issues in the United States can be greater among individuals with more education and science knowledge" (Drummond and Fischhoff 2017: 9590). Evans and Durant (1995) have shown that "while knowledgeable members of the public are more favorably disposed towards science in general, they are less supportive of morally contentious areas of research than are those who are less knowledgeable" (p. 57). Human embryology is an instance of a morally contentious issue: as a consequence, scientific literacy is not always positively correlated to a pro-attitude toward science, or to the belief that science can create more favorable conditions for our life. Analogous conclusions are reached by Allum et. al. (2008): a high level of scientific literacy is not always accompanied by a positive attitude toward science in specific domains of its application (in biotechnology, for example).[16]

Admittedly, the scientifically literates' perception of a conflict between political and religious ideas and scientific evidence can be dissolved only in certain cases. On the one hand, scientific education can assuage even deep commitment to political and social interests: citizens in good faith coming to learn solid scientific evidence about a rapidly increasing global warming will change their minds even in the presence of strong economic interest. Jehovah's Witnesses' prohibition of blood transfusions can be overcome by carefully transmitted scientific information. The fact that some member of the sect will stubbornly remain convinced that transfusions are sinful is not evidence against the importance of a greater scientific literacy. These cases, as for the most fanatic flat-earthers, must be studied carefully since they cannot just be classified as forms of madness or paranoia. They cannot be tackled here exactly for this reason.

On the other hand, historical studies show that it is more difficult to have politically interested literate citizens or even scientists admit their lies about scientific evidence concerning, say, climate change. These days, the deliberate creation of unwarranted doubts about the anthropogenetic influence of climate change may irreparably delay the necessary interventions to save our planet. This fact shows that scientific literacy may be insufficient for an effective intervention toward bettering public health or even guaranteeing a future life for the next generations despite the fact that it is necessary.

Ethical stances are more delicate, since they can remain stable independently of the level of scientific education. Abortion is an example, but to the extent that a very good level of scientific education leads to atheism, religious attitudes against the interruption of maternity or the end of life can be reconsidered. Likewise for questions related to embryology, that are usually influenced by spiritualistic philosophies that must of course be respected.

As far as technology is concerned, a lot of educational effort must be put into fighting against the illegitimate identification of science with technologies exploiting the resources of the planet and creating new weapons. In addition, there are objective difficulties of adaptation towards a technology that develops faster and faster. Biological adaptation to the environment has been *slow* and has had a very long history, but the cultural environment changes much *faster*. The so-called "infosphere," as well as robotics with artificial intelligence, are making many human works and activities obsolete, causing unemployment, and requiring very rapid adaptation times that make long-term social predictions and therefore planning impossible. These phenomena require some kind of life-long education.

6.3.2 The Presence of Functional Illiteracy Also in Developed Countries

Besides a commitment to ethical, economic and religious issues, negative or less than enthusiastic *attitudes* toward technology and scientific applications (even in scientifically literate people) can be explained by additional causes. Together with scientific disinformation, with the dire political consequences that we have seen, one neglected factor contributing to the spread of web-generated misinformation, is the so-called *functional illiteracy*, a phenomenon that is present in all developed countries, European countries included. In a recent report, we read: "According to the recent literacy rate, 85% of the adult population in the world is literate, and therefore worldwide about

757 million people are illiterate (UNESCO 2015). Large-scale assessments in developing countries indicate that illiteracy is more prevalent, while in developed countries, functional illiteracy is prevalent,"[17] which is a sort of relapse into illiteracy. "Literacy is defined as the ability to understand, evaluate, use, and engage with written texts to participate in society, achieve one's goals, and develop one's knowledge and potential" (OECD 2013: 59). Functional illiteracy is usually defined as the inability to effectively understand a text, or to write correctly, or to apply mathematical skills useful for everyday life, with the consequent inability to use printed and written information to function in society.

Reports from the Program for International Student Assessment (PISA) concerning the ability of 15-year-olds to use their reading, mathematics, and science knowledge show some worrisome data. On average, 17 percent of 15-year-old Europeans have poor reading skills and cannot understand their school textbooks, while 18 percent of 9-year-old children almost never read a book outside school.[18]

The data for European adults are not more encouraging. According to the EU High Level Group of Experts on Literacy, in 2012, "an estimated 20% of all adults in Europe is probable to [sic] lack the literacy skills needed to function fully in modern society."[19] This means that an estimated 73 million European adults lack "functional literacy," which implies that even though these adults can understand simple texts, they are not able to deal with longer or more complex texts by interpreting them in a correct way. In addition, these data are correlated with increasing difficulties to find jobs and to avoid the risks of poverty and social exclusion, a fact that implies the necessity of life-long learning. Similar data are confirmed by the Program for the International Assessment of Adult Competencies (PIAAC).[20]

Given these numbers of functionally illiterate individuals, it is not surprising that even in countries that must be considered to all intents and purposes economically developed there is a substantial part of the citizens who are subject to manipulation essentially due to the web. This is the fundamental fact explaining the spread of fake news via the internet, according to which, for example, immigrants live in large hotels while earthquake victims live in shacks. It can also account for irrational attitudes towards well-tested consequences of the assumption of drugs, accompanied by the adoption of homeopathic therapies even in cases of serious illness and for the hatred toward expertise and the tendency toward populism and direct democracies. Functionally illiterate citizens can more easily be manipulated.

A low level of education also has important economic repercussions, given that it is statistically correlated to crime and the poverty index. For example, a good percentage of prisoners are functionally illiterate, since they were born in degraded and poor social environments where criminality prevails and in which it is very difficult to enjoy higher education. In these environments, the underworld prevails, and its presence creates further poverty in a vicious circle, given that the presence of criminal organizations discourage investments. In general, in a country in which the difference in wealth is more marked, there are also greater difficulties in guaranteeing all young people a good average level of education. The school thus loses its function of being a "social elevator" precisely for those to whom the social lottery has reserved a worse fate, namely the poor. The loss of this function, combined with the low salary of teachers who are sometimes not well selected, has made teachers lose a lot of respect and social dignity, although their role for the future of society is decisive.

Note the vicious circle, which perpetuates the social imbalance between rich and poor areas of economically developed countries. It can be safely assumed that the low cultural level of the family of origin is almost always due to a condition of greater poverty, and it is precisely this fact that does not lead to reading books or newspapers, even in digital format, but rather leads to gathering news from often unreliable social networks. In their turn, it is due to these factors that the social inequalities are perpetuated, because the low cultural level is responsible for the difficulties in improving one's social status.

These data obviously imply that scientific literacy goes hand in hand with general literacy. There are empirical studies that support this plausible claim by stressing the important role that reading and writing skills have in developing scientific literacy (Feinstein 2010: 172; Norris and Phillips 2003). More in general, and conversely, this link suggests that a well-rounded education does not only entail being curious toward literature, the history of art, and philosophy, but also to be able to orient oneself, for example, on the rudimental principles of modern and contemporary physics and, more relevant for scientific literacy as I understand it here, on the principles of genetics and evolutionary biology. The former principles teach us that the universe is approximately 13.7 billion-year-old and is evolving, that only in our galaxies are there *c*. 100 billion stars and that therefore we occupy a minuscule place in the universe. The latter principles are more relevant to understanding the origin of ethical values and therefore their place in our society, while genetics is essential in questions related to inheritance, cloning, etc.

An obvious objection is that this form of "theoretical knowledge" is not immediately relevant to solve daily problems or problems of social interest. However, as Charles Snow ([1959] 2013) convincingly argued, the gap between the "two cultures," the humanistic and the scientific one—which, given the minore did not exist between the time of the ancient Greeks and the Renaissance given the minor specialization of knowledge—must be at least partially bridged. According to Snow, who was a writer and not a scientist, the polarization of the two cultures makes it much more difficult to solve the problems of the world. While society is increasingly influenced by technological developments, the humanistic culture in general is indispensable for governing them by considering the essential contribution of art in general to the formation of an ethical spirit. History is indispensable to locate the contemporary social problems in a wider perspective. Shen's notion of Cultural Scientific Literacy expresses Snow's idea: "cultural literacy [It] is motivated by a desire to know something about science as a major human achievement [...]. It solves no practical problems directly but it does help bridge the widening gulf between the scientific and humanistic cultures" (1975: 267).

To this end, the publication of good popular essays in any scientific field is a civil service that the best scientists and competent journalists should accept more often for the sake of the community. Popularization of science would make it easier to transmit correct scientific information not only in schools of all kinds and levels, but also among the public at large. Scientific popularization implies a *translation* between different languages, which is beneficial also to science. Only someone who has a deep knowledge of a scientific theory is able to explain it also to laypeople and to scientists working in different fields. Explaining difficult scientific concepts by using ordinary language (which is obviously largely different from the scientific jargon) helps teachers and scientists to better understand their own technical jargon. The better they understand their field of expertise, the more are they capable of explaining it to 12-year-old children.

A higher level of scientific literacy would make manipulation of opinion due to disinformation much less likely to take place. It would be the best solution to one of the most pressing problems of our society, namely discriminating between the pseudo-experts and the real experts. This discrimination is the only way to make possible a rational choice between two scientific policies that have social consequences but that conflict with each other. Still with this in mind, I will end the chapter by suggesting a criterion allowing us to evaluate at least approximately the competence of an expert quickly and easily. This criterion, while involving the use of very little cognitive effort, is at the same time potentially effective, because it just requires knowledge of some of the *social* mechanisms allowing the selection of scientific hypotheses. Its validity therefore depends on what has been argued in Chapter 2. Perhaps paradoxically, it entails the capacity to make use of an instrument that very often conveys unfounded or false beliefs: the web!

6.4 The Evaluation of the Expert's Competence Through the Web

The radical condemnation of, and unlimited enthusiasm for, the web represent two opposite and equally unjustified attitudes: Umberto Eco was correct when he wrote that social media gives the right to speak to legions of idiots: "the TV had promoted the village idiot to someone to whom the viewer could feel superior. The drama of the Internet is that it promoted the village idiot to the bearer of truth."[21] This type of phenomenon is favored by our natural desire to be able to feel superior to people who are more famous because they can very often be seen or heard in the media and in talk shows.

As with weapons, the more powerful the means that are used to achieve a certain purpose, the more important it is to know how they are to be used. An *approximated* rough-and-ready evaluation of the scientific quality of the protagonists of a debate that sees real experts on one side and charlatans on the other ought to use the same procedures of selection that are used in science. Scientists who are competent in field A and want to have an idea about the competence of colleagues working in a very distant field B usually rely on the web.

In these cases, consulting only the curriculum vitae of the protagonists of the debate is in many cases sufficient, even if some sites may not be completely reliable. But one can check different sites, as happens when we read different newspapers. For instance, the fact that self-styled experts, unlike their opponents, have no position either in the academy or in research institutions should already raise our suspicions. It is easy to go to a university website and consult a scientist's membership in a department or research center of a national or international university. Becoming aware of the main social criteria to evaluate the competence of a supporter of a given position in a public debate helps also the non-specialist to realize that there is an important difference between a self-styled expert and someone who has earned a degree from a good national or international university. This does not mean that the latter cannot be wrong and the former right, nor that CVs cannot be altered, but on average these cases are quite rare.

Some scientists use databases and sites to help them to establish the reputation of certain of their colleagues through criteria that evaluate not only the quantity but also the quality of publications. Publicly accessible sites of this type for scientists are represented by databases such as Scopus or Web of Science, which intend to measure, albeit controversially, the impact and interest that the publications of a certain scientist or a team of scientists have had in the relevant field of investigation. Could not laypeople use these criteria as well?

The questionable nature of these databases depends on the fact that purely quantitative aspects are not indicative of the research quality. For instance, there are quantitative indexes such as the so-called h-index created by the mathematician Jorge Hirsch: an h-index of 100 means that 100 publications by a scientist have obtained at least 100 citations. The problem is that a scholar P can agree with a scholar Q to cite the other's work and conversely, so that the number of citations of the two scientists can grow without generating a significant progress in the field of investigation but only increasing artificially the reputation of the two scientists. However, there are methods that sifts these data and remove them from the evaluation. Sometimes a single article can have much more influence than many articles and many articles are signed by hundreds of people, so that the number of citations grows.

There are ways also to measure these facts that here I will not discuss since they have been the subject to a very wide debate. Suffice it to say that in their attempt to *precisely quantify* the scientist's authority, bibliometric criteria of this type are naive: the scientist's competence is instead the fruit of well-founded collective judgment by the relevant scientific community. However, If scientists of group A have published even a few articles in highly authoritative journals, it is more than legitimate to assume that their work has a much more solid scientific basis than that of most 'experts" in group B. This holds in particular when the latter have, if at all, written articles or books that appeared in magazines and publishing houses with a purely local circulation, or who have not even written anything except articles on the web that have not passed any critical scrutiny. Scientists use these criteria all the time. With little time available, even non-expert citizens who are aware of the evaluation mechanisms of a scientific contribution can quickly discover that only some protagonists of the debate have published a book with authoritative national and international publishing houses. To work properly, these mechanisms should of course become widely known to a large public of non-specialists, a fact that does not seem impossible First of all, the cv of the debaters is publicly avaliable in the web. Second, research published by group A in international journals is typically.

Typically subject to those very strict selection criteria already presented in the second chapter, If those who disagree with them have published papers in journals with no international impact, it is not difficult to choose the "right group" of competent scientists.

Sometimes we hear it said, even in the academic world, that a lower-level paper can be published also in journals with an excellent reputation while very good ones are rejected because of the existence of certain schools favoring given methodologies instead of others. However, in view of the socially based selective criterion already discussed above, *statistically* this happens very rarely and good journals are usually open to good papers that are not mainstream. Furthermore, it is rarely the case that a very good scholar risks her reputation by submitting her paper to journals with a low international impact. As mentioned before, contrary to journals which decide to publish very few articles sent for evaluation, journals of much less importance accept almost all submitted papers. Moreover, if an author believes that one of their articles is very good and original, and the public to which it is addressed is international, they will try to publish it in international journals that are read by all.

Considering the level of specialization of scientific knowledge, the same argument also applies in cases where the dissent is among scientists and not among scientists and charlatans. An expert in geological or climatological fields is more authoritative on the issue of global warming than a scientist who has worked only in particle physics.

To summarize, an assessment of the scientific authority of "conflicting experts" based on the quality of their publications can be carried out also by laypeople and it is a good guide to side with the most reliable hypotheses. On this point, the educational responsibilities of the local media are important, given that the best academic institutions have their product evaluated on the basis of universalistic and supranational criteria of selection. Just to refer to two cases that have had a wide echo in Italian newspapers, an online check of the quality of Di Bella's or Vannoni's scientific productivity (see p. 90) would have been enough to unmask their deception. Having the highest number of followers on the web and being cited in international scientific journals are two completely different matters.

7

The Role of the History and Philosophy of Science in the Democratic Debate

In this chapter, I will show why integrating scientific with humanistic education is not only one of the most powerful tools to strengthen the scientific literacy process but is also indispensable for further defending representative forms of democracy. The inevitable "analytic" fragmentation of technical-scientific knowledge is in fact an element that pushes citizens to claim greater autonomy of choice on issues that concern them closely but on which they must rely completely on others. In order to reconcile the competence of experts and the autonomy of citizens practical and theoretical, the history and philosophy of science can play an essential role. It is thanks to them that the multiple perspectives offered by the individual sciences do not make us lose sight of a more "synthetic" look at human knowledge. As we will see, these two disciplines are essential for highlighting the typical characteristics contributing to the formation of a scientific attitude, which is indispensable both for defending oneself from widespread misinformation and for making informed decisions on issues of science policy.

To defend the role of the history and philosophy of science in the training of scientists, Einstein once made a similar point. In 1944, he wrote a letter to a young physicist who wanted his support to introduce "as much philosophy of science as possible" into the physics department where he was preparing to teach. Einstein replied:

> I completely agree with you on the importance and educational value of methodology and the history and philosophy of science. Many today – and even professional scientists – seem to me like those who have seen thousands of trees but have never seen a forest. Knowledge of the historical and philosophical background provides the kind of independence from the

prejudices of one's generation that many scientists lack. The independence created by philosophical understanding, in my opinion, is the distinctive trait that separates a mere craftsman or specialist from those who really seek the truth.

Einstein 1986: letter to Thornton, December 7, 1944,
EA 61-574

To defend Einstein's thesis, in section 7.1, I will begin with some considerations on the role that the history of science has in the cultural formation understood in its broadest possible sense. The history of science enables us to understand more than other disciplines the change in the cognitive modalities with which we have investigated the world surrounding us. Thanks to the history and philosophy of science, not only have we learnt directly about the natural and social world, but also about *how* our knowledge thereof has developed. In section 7.2, I will highlight three reasons why the philosophy of science can play a pivotal role in helping to raise the level of scientific literacy. The first is its naturally critical and traditionally universalizing function. The second involves some characteristics of the scientific method intervening in the evaluation of any scientific hypothesis. The third function depends on the fact that it is the best weapon against any form of cognitive relativism, according to which all scientific claims to truth or epistemic criteria are relative to historical, geographical, and social contexts, so that there is no true, correct, or mistaken point of view on any scientific description of the world, and therefore also on any stance in public debates involving science.

7.1 The Role of the History of Science in the Formation of a Scientific Mentality

The history of science is fundamental to achieving a deeper understanding of the evolution of the fundamental concepts on which science is based. It can be studied by privileging a point of view that is more *internal* to theories—and therefore more attentive to their transition from one set of hypotheses to the other—or more *external*, and therefore more sensitive to the interactions between science and the institutions that have facilitated its development (think again of the role of academies in modern times).[1] In both cases, it must be considered as a bridge between the scientific disciplines and the humanities and therefore as an indispensable element to limit the negative effects of hyper-specialization. Of course, historians of science are also specialists. However, by placing scientific theories in a wider historical framework, their discipline is an essential tool for bringing together the humanistic and scientific cultures, Snow's two cultures that mentioned in the previous chapter. For example, a book on the history of technology and its social consequences would help us to better understand many contemporary phenomena.

I will present very briefly two reasons, linked together, which motivate this claim. The first (7.1.1) defends the claim that only the history of science can make us fully appreciate the progressive distancing from anthropocentric visions of the universe by informing about our place in it. The second (7.1.2) deals with the choice between two opposite theses on the nature of scientific change: (a) the path of science proceeds without discontinuity, or (b) science advances via revolutions, with important implications for assessing the credibility of a new hypothesis in a public debate involving scientific knowledge. In the second chapter, I mentioned that even the most innovative theories must take into account what has already been discovered and can never be denied in its entirety. This is an important argument against the thesis that scientific truth changes radically from historical moment to historical moment and that there is no correct point of view on public issues on which we are called to decide.

7.1.1 The Overcoming of Common Sense by Science

The most important cultural aspect of science consists in its liberating power towards anthropomorphic or anthropocentric visions of the world dictated by common sense. It is precisely on this aspect that the education to scientific thought of the new generations should focus. As examples, we can think of the troubled history of the discovery of the Earth's motion, of Darwin's evolutionary theory, the discoveries of the neurosciences, or to the fact that, according to the theory of relativity, time "flows" differently, depending on the spatiotemporal trajectories described by clocks and their position in the gravitational field. Last but not least are the bizarre aspects of the quantum world, where our language, for the majority of physicists, is not applicable: to claim that a quantum entity is both a wave and a particle implies that it is neither.

For reasons that I have already presented in section 4.3 and that will be further defended in the rest of this chapter, all of these disciplines describe objective aspects of our scientific image of the world, which are in sharp contrast with our manifest image,[2] and as such are counterintuitive: as Carlo Rovelli put it, "reality is not – always – as it appears."[3] It is only thanks to the history of scientific theories that can we appreciate how long and tortuous the process has been which has led to the overcoming of common-sensical, implicitly taken-for-granted hypotheses that made the greatest scientific revolutions more difficult to achieve. One example, which will only apparently lead us out of our path, will suffice to illustrate this thesis.

In the modern world, a gigantic step forward toward our understanding of nature concerns the following remarkable fact: between a body that *moves* with uniform rectilinear motion (i.e., that continues to move without accelerating) and a body *at rest*, there is no physical difference. This is Newton's first law of mechanics, the law of inertia. The reason why those who study it for the first time consider it obvious or even trivial is because they are unaware of the efforts made by various generations of philosophers and scientists to understand the nature of motion. In Aristotle's physics, there is an absolute difference between rest and any type of motion, and any motion presupposes a mover. In his terrestrial physics, the natural motions of bodies are either upwards (air and fire), or downwards (water and earth).[4] Being "natural" in this sense, they do not need an explanation. One of the main questions in Aristotelian physics is how to explain the non-natural or "violent" motions, as he called them, which are mainly caused by human beings, and which are neither upwards nor downwards. If there is nothing that apparently pushes a stone after it has been thrown by our hand, why does it keep moving for a while until it touches the ground?

Historically and in a very schematic way, two competing theories were advanced to explain this type of motion: (i) a flying arrow generates a circular motion of the air that keeps it moving by pushing forward its tail, and (ii) the air receives from the original mover (the hand) the power to act in turn as a mover, or to push the air in front of the arrow and the arrow itself (Clagett 1959: 533).[5] Interestingly, these two theories are presupposed by our "common sense," as evidenced by various experiments performed on students interviewed by psychologists who try to understand the naive physics inscribed in our brain (McCloskey, Caramazza, and Gree 1980). The erroneous assumption of common sense that Aristotle made his own is that *any type of motion* requires the presence of a mover (a "force" in an anachronistic language) that continues to act in contact with the body.

This historical fact helps us to understand why postulating the law of inertia required the remarkable capacity for abstraction of geniuses like Galilei (1564–1642), René Descartes (1596–1650), and Newton. Thanks to these natural philosophers, we now know that a body moves inertially if, and only if, no force acts on it. So, we have to eliminate in thought the presence of the force of gravity and friction while the world in which we live is dominated by these two forces. Thanks to the fact that (a) there is no physical distinction between a body at rest and a body moving rectilinearly with uniform speed and (b) in short intervals of its orbit (neglecting its rotation), the motion of the Earth is approximately inertial, we can explain why our planet moves without us being aware of it.[6]

Likewise, coming to learn that the complexity and the goal-oriented character of a human eye is essentially due to a Darwinian, very long process of blind genetic variation and natural selection goes against more anthropomorphic explanations that invoke either a Lamarckian transmission of traits acquired during the animal life[7] or, more often religious accounts. The same holds for our coming to learn the nature of genetic mechanics and the fact that mental states are either identical with, or produced by, very complex dances of parts of the brain.

An appreciation of the historical dimension of science enables us to understand its cognitive significance, and the liberating nature of science with respect to these "commonsensical beliefs or appearances" is destined to make the process of understanding science more compelling and fascinating, both in schools and in the wider public, with the predictable results of raising the appreciation of science and therefore also the level of scientific literacy. A greater emphasis on an historical teaching of scientific subjects should not, however, be limited to secondary schools. Following Einstein's previous quotation, even the university training of a young physicist, for example, should not only be based on rote or "mechanical" learning of laws, formulas, and nomenclatures. Knowing how to locate a scientific problem in its historical framework means understanding it in more depth and therefore making the solution of new problems more probable. Realizing this fact is enough to create distrust of anthropomorphic, anthropocentric, or mythical explanations of natural phenomena, which are not infrequently circulated as if they were comparable to scientific ones.[8]

In a word Understanding the fact that revolutionary changes in our worldview have been made possible by the overthrow of anthropomorphic theories implicit in common sense would certainly be of greater interest to budding students, contribute to generate greater literacy in science, and guide citizens in making more informed choices in public decisions involving naive, prescientific views of nature and our place in it.

7.1.2 The Incommensurability of Scientific Change and the Acceptance of Theories Alternative to Official Ones

However, the awareness, which developed thanks to the great revolutions of the twentieth century, that science can radically change its conceptual framework generates the problem of truth of its theories and therefore, in intellectual circles, of trust in its practitioners.

On the one hand, scientific revolutions strengthen the idea of the fallibility of science, discussed above.[9] The reader may recall from the second chapter that in the course of the history of science even the best theories have been superseded, with the result showing that they are "false" in some respect, that is, relatively to some values of the relevant magnitudes in given fields of their application. These partial refutations are exactly what has made possible the formulation of theories that are experimentally more accurate with respect to the previous ones.

On the other hand, the thesis that scientific change can often be radical has been used to argue that every scientific revolution radically modifies all basic scientific hypotheses accepted by a research community. By using this historical argument speciously, an attempt has been made to justify a methodological principle, according to which scientific progress is made possible by a proliferation of methodologies and hypotheses which are *totally alternative* to those of the official science (medicine included), with respect to which the latter, however, should assume an attitude that is at least humbler. In the end, all methods of acquiring knowledge, whether scientific *or not*, ought to be regarded as being on the same level. As anticipated above (p.61) To express this anarchistic idea about the knowledge the Viennese philosopher of science Paul Feyerabend (1924–94) claimed that "anything goes" (Feyerabend 1975).

It is therefore no coincidence that Feyerabend, in a way more radical than Kuhn, argued that the transition from one physical theory to another involved a change in the *meaning* of fundamental physical notion—such as, for example, space, time, and mass—which, in turn, involved an *incommensurability* between pre-revolutionary and post-revolutionary theories. The meaning of these fundamental notions, that in physics are essential, we are told, depends holistically on the whole language of which they are a part.

This is an argument often used by Feyerabend to argue that scientists have now become priests of a new faith, intolerant of alternative world views. For this reason, he invited us to be wary of experts (and therefore also of "traditional doctors") because, as he wrote, they are "full of prejudices, they cannot be trusted, and their advice must be thoroughly evaluated" (Feyerabend 1978: chapter 5, especially part 2). As we have seen in the previous chapters, this claim is justified in part by historical episodes involving dishonest scientists.

Feyerabend's controversial pluralism questioning the existence of *a* unique method of science (Oreskes, 2019, shares this view), and his invitation to be prudent about the hypotheses of experts, however, was accompanied by a radical and unjustified attack on the evidential strength of their theories. In many social issues, he maintained, one had to replace the experts' advice with *citizen's initiatives*, an anarchistic view of scientific knowledge that in his opinion justifies some form of direct democracy. Independently of the correctness of Feyerabend's view about science and expertise, what matters for my main claim *is the correct link that he established between the epistemology of science and different forms of democracy.*

It is easy to imagine why such a relativistic conception of human knowledge lends itself to exploitations against the principle of the objectivity and neutrality of scientific knowledge and can be used in favor of the idea that the affirmation of one theory over another is only possible thanks to forms of latent propaganda arising from social, political, and economic factors. Abandoning the high-level intellectual sphere thanks to some forms of popularizations, this argument can be used in a conspiratorial way, thereby calling into question powerful lobbies, which would bend one theory to the detriment of others. By radicalizing it, this form of relativism lends itself to argue that there is no real difference between experts and incompetents.

As a matter of fact, even a profound change in the meaning of some fundamental terms such as "space," "time," and "mass" does not imply that the achievements of previous theories have been lost.[10] Contrary to what Kuhn and Feyerabend argued and despite the diversity of their perspectives, the second chapter showed why the *wave* theory of light and Einstein's theory of general relativity were a generalization, respectively, of the previous *corpuscular* theory of light and of the *Newtonian* theory of gravitation. As explained in that chapter, the term "generalization" means that the latter theories hold only in particular circumstances that is, under appropriate, more limited conditions. Given the complexity of the issue, these two examples cannot be considered to be definite arguments in favor of the "continuist" conception of scientific change. However, they help me to illustrate two facts that are particularly relevant:

1. The two philosophical positions on the nature of *scientific* change are reflected in two ways of understanding *political* change. The "continuist" holds that there is scientific progress because even when there is a radical scientific transition, the main successes of the previous theory are kept also by the successor theory. The "discontinuist" theory of scientific change instead holds that when there is a conceptual revolution, the later theory becomes incommensurable with the earlier one: the two theories use a different, non-translatable languages. The continuist conception of science is mirrored by political views that stress the importance of *gradually improving* already existing social conditions without destroying them. The "discontinuist" conception of scientific change can be likened to the view that any political change in a democratic society can only be possible when the main foundations of the previous social system are overthrown. No change is possible by "generalizing or extending" previous political systems. It is not a coincidence that Kuhn's *Structure of Scientific Revolutions* gained attention also because of the term "revolution" that appeared in its title during a period (early 1960s) in which pressures toward radical societal changes were mounting.

2. The second fact is that scientific continuism is linked to the consistency of the new hypotheses with respect to the old ones: a new scientific hypothesis is accepted by scientific communities only if it does not contradict most of the beliefs already acquired and taken for granted, the "old" hypotheses. This principle of *consistency* can hardly be underestimated. In fact, it is the compass that scientists always hold in their hands to distinguish a theory proposed by a competent scientist from a hypothesis put forth by an incompetent. There cannot be wholly new scientific theories, even though the progress of science depends on new, hitherto non-considered hypotheses.

The principle of consistency has an implicit role also in many situations of everyday life: if we were told that an event that we consider very unlikely in light of everything we know has happened, then we would and should have many doubts about it. We have seen why what we know does not and cannot depend only on our direct experience, but on that of countless other human beings. Exactly as scientists do, we must always assess a new hypothesis by comparing it with the beliefs that we have already acquired and that belong to our background knowledge.

In the case of scientific communities, a new hypothesis can pass an initial test of plausibility and can be later pursued if and only if it does not contradict many other beliefs already experimentally confirmed several times. What competent scientists do not question because they consider it very probable should not be doubted by the rest of us, scientists working in different fields included. It follows that the novelty of a new hypothesis and its compatibility with all of the others must go hand in hand. Ultimately, scientific literacy and the consequent ability to take a position that is as informed as possible in order to make a decision with ample social relevance largely depends on the public availability of this background scientific knowledge, even if not specialized. *It is this knowledge that allows us to evaluate the coherence and plausibility of our beliefs with a method that is structurally identical to that used by specialists.*

The broader meaning that this italicized statement has from our point of view must not be overlooked. To accept the beliefs of other experts by appropriating them in a coherentist framework (see next page) is not incompatible with a scientist's independence of judgment with respect to intellectual influences that in some cases could stifle her research program.[11] A similar principle applies to the citizens of a representative democracy who, by becoming aware of this coherentist epistemology,[12] can reconcile the autonomy of their decisions both with the need to trust the experts and with the task of finding the most effective and morally acceptable means to achieve their individual ends.

To sum up, in this section we have seen why even a non-specialist knowledge of the history of scientific ideas is a very important weapon that can be used to defend ourselves from the prevailing misinformation. Becoming aware of how the greatest scientific revolutions have led to the abandonment of anthropomorphic visions of reality can have important consequences. In a public debate in which anthropomorphic theories like telepathy or telekinesis were defended in one part of the camp, scientifically equipped citizens would immediately know which side to take.

7.2 The Role of Philosophy of Science in the Public Debate

In the seventeenth century, what we now call "physics" was referred to as "natural philosophy."[13] With some caveats, the interaction between the philosophical foundations of scientific disciplines and the actual sciences has remained fruitful to this day (Huggett 2010). In the first decades of the twentieth century, the founding fathers of the new quantum mechanics—besides Einstein, Werner Heisenberg (1901–76), Niels Bohr (1885–1962), Max Born (1882–1970), Erwin Schrödinger (1887–1961), Hermann Weyl (1885–1955), among others—were deeply interested in the philosophical consequences and conceptual foundations of the new revolutionary theory that they were contributing to create. In this sense, they actively defended the importance of a non-superficial interaction between physics and philosophy.

Partly due to the Second World War and the commitment of scientists in the construction of the atomic bomb, this interest was lost at the turn of the last century. In the last forty years or so, however, a fruitful interchange between sciences and the philosophy of nature has regained vigor, also thanks to the fact that a new generation of philosophers more specialized in the individual sciences has competently mastered the related language of the latter. This does not imply that the philosophy of the sciences and the sciences must lose their autonomy (see Dorato 2011), but only that between scientists and philosophers of science there is currently a division of labor bringing them culturally closer than they were in the middle of last century. While scientists are generally more attentive to the relationship between evidence and hypotheses, philosophers deal with the implications and conceptual foundations of the involved theory.

For example: what is the relationship between the concept of space (or time) presupposed by physical theories and that of our experience of the world? While scientists work specializing separately in physical and psychological space (physicists and psychologists, respectively), philosophers focus instead on the nature of their relationship.

In what follows, I will briefly discuss three reasons why a familiarity with the philosophy and methodology of science can help to orient oneself in debates concerning scientific policies despite the inevitable lack of specialistic knowledge. The first (7.2.1) refers to the fact highlighted by Einstein in the above-mentioned quotation. The philosophy of science can provide more general knowledge than the individual sciences through its "synthetic vocation" which consists in placing individual scientific theories and debates into a broader cultural context. The second reason (7.2.2) explains in more detail how it can realize this vocation: in evaluating the plausibility of scientific hypotheses philosophers of science inquiry into notions that are fundamental in all scientific discipline, namely, *explanation, probability, causation,* and *logical consistency*. Finally, the third reason (7.2.3) discussed at the end of this section is given by the fact that philosophy is the best weapon to combat cognitive relativism. Following in the footsteps of a certain way of reading the above-mentioned Kuhn and Feyerabend's philosophy, for epistemic relativists there are no facts independent of the conceptual schemes that serve to organize them and which we could compare to biscuit molds that give shape to an amorphous dough (the facts).[14]

The elementary functioning of these three critical functions can be taught in high school and yet is very important for developing an awareness of the methodological and cognitive procedures of science. In the last 150 years or so, the philosophy of science has undergone a process of progressive specialization and theoretical consolidation not unlike that of science. It has therefore become more and more important to incorporate the point of view of the philosophers and sociologists of science precisely on the relationship between scientific knowledge and its applications, given that it has been the subject of many debates and critical discussions entrusted to international journals, papers, and conferences, just as happens in the most common, properly scientific knowledge.

7.2.1 The Importance of the Cultural and Methodological Synthesis Offered by the Philosophy of Science

To argue that the *practical utility* of the philosophy of science lies in its "universal" vocation is certainly paradoxical. This universality can be interpreted in a double, apparently contradictory, way. On the one hand, it indicates that the philosophy of science, like the philosophy of other branches of knowledge, or like art, is an end in itself, in the sense that it has a purely cultural meaning: it is not a means to any given end. On the other hand, and precisely because of its universality, the philosophy of science has important repercussions for our discussion. It is the second meaning here that is relevant.

To clarify the universal meaning of the philosophy of science we can go back to Plato, who had already claimed that the dialectic (the philosopher) must be capable of a "synopsis," that is, of seeing all ideas with a single glance and therefore of grasping unity in what is multiple and different. The individual sciences, on the contrary, give a partial image of the world: the inanimate "perspective" (physics and chemistry), the living perspective (biology), the mental perspective (the cognitive sciences), the social perspective (history, sociology, and economics), etc. On the contrary, the philosophy of science has the task of inquiring into the possibility of putting together these variegated pieces of the puzzle in order to provide an image that is as unitary as possible of our place in the universe. This image, even if it existed, could never be considered ultimate because it is fallible, revisable, and improvable, just like that of the individual sciences composing it.

From a practical point of view, the ways in which the philosophy of science works to realize its unifying vocation has often consisted in asking whether, beyond their different objects of investigation, all the natural sciences (and even all the empirical sciences, including social sciences) share the same method of hypothesis-testing. In my opinion, "the greater generality" of the philosophy of science does not necessarily depend on the existence of a unique method. Rather, it depends on the fact that it has always critically discussed and continues to analyze in ever more rigorous ways the concepts that ground *all* the natural sciences, such as probability, causality, consistency, and confirmation of hypotheses.

As we shall see by discussing some examples, a deeper understanding of the meaning and role of these concepts is essential to intuitively evaluate the plausibility of the hypotheses raised in public discussions involving scientific questions. In its task of analyzing key conceptual notions, the philosophy of science asks questions like: What does it mean to claim that a *hypothesis is reliable or probable*? When can we believe that one event is the *cause* of another? What does it mean to *explain* a phenomenon and when can we assume that a hypothesis is *confirmed* by the facts? From this point of view, the philosopher can be seen as an *engineer working with concepts* and philosophy as a form of "conceptual engineering" (this definition is due to Floridi 2014). Philosophers of science work with the scientific concepts listed above.

To illustrate this view, a typical way in which the philosophers of science carry out "conceptual engineering" consists in elaborating the so-called *explications* (Carnap 1950). These consist in: (i) beginning one's analysis from the vague meaning of concepts such as those reported above in italics, to which, however, an intuitive sense is associated, (ii) trying to make this meaning as clear and rigorous as possible, and (iii) keep as much as possible some central senses associated with the original concepts: "[t]he task of explication consists in transforming a given more or less inexact concept into an exact one [...]. We call the given concept (or the term used for it) the explicandum, and the exact concept proposed to take the place of the first (or the term proposed for it) the explicatum" (ibid.: 3).[15]

7.2.2 The Role of Probablility in the Formation of a Critical Attitude Toward Fake News

Within a unitary conception of our knowledge, the formation of a scientific way of thinking requires an elementary knowledge of the theory of probability that can be already acquired at secondary school. Probability, as the great economist of the last century John M. Keynes (1883–1946) put it, "is the guide of our life." In the same vein, in his 1951 presidential address to the American Statistical Association, the mathematical statistician Samuel S. Wilks (1906–64) paraphrased a quotation from H. G. Wells (1866–1946): "statistical thinking will one day be indispensable for being responsible citizens as the ability to read and write" (Wilks 1951: 14).[16] In fact, probability plays a fundamental role in all sciences, from quantum physics to genetics, and in all social sciences, particularly in economics and sociology.

Also, in everyday life, we unconsciously orient ourselves in the world by assigning a subjective degree of belief to a certain event. The already mentioned eminent founders of the subjective conception of probability—the English economist mathematician and philosopher Ramsey and the Italian mathematician de Finetti—argued that we could measure our personal degree of belief in the happening of a certain event on the basis of the sum that we would be willing to lose in a fair bet. For instance, would you be willing to lose €9 (or 9 times the stake) and earn only €1 by betting on the victory of a football team? Since in your bet the defeat is subjectively estimated 9 times more likely than the victory (9 to 1), you assign a very low probability to the event of the victory (¹⁄₁₀ or 10%). If the amount you would be willing to lose were even lower, the probability that you would assign the victory would be proportionally lower. This type of probability is very important for evaluating the plausibility, even subjective, of a hypothesis in light of what we know and our personal background knowledge. This subjectivity can however be overcome in all cases in which the personal degrees of probability of many people converge to the same number, or when we are dealing with objective probabilities or *chances*, whose probabilities do not depend on our variable degrees of ignorance. In the latter case, the rule dictates that our subjective degrees of beliefs should become identical to the value assigned to chance (Lewis 1980).

Among the objectivistic theories of probability, I have already introduced one that is highly relevant in our context, the frequentist theory of probability, that here needs to be briefly rehearsed.[17] A typical case of scientific misinformation defends the harmfulness, say, of a particular vaccine on the basis of its serious contraindications. In cases like this, knowledge of the frequentist conception of probability proves to be of fundamental importance since the variable estimates of the contraindications of all vaccines would not be ignored. Suppose that a vaccine against hepatitis B causes a serious allergic reaction once in every 600,000 administrations, which is the corresponding frequency. But also suppose that in untreated, acute forms, this disease caused in the unvaccinated population 500 deaths in every 10,000 cases.[18] After the vaccine, the probability of contracting hepatitis is not null but becomes very low (0,00000167), whereas without the vaccine it is much higher (0.05). Therefore, the fact that the vaccine may have contraindications does not constitute a good reason for not being vaccinated, given that the event in question is very unlikely and it has many more advantages than disadvantages. For the same reason, not wanting to risk anything, we should not even take an aspirin, as it could cause an allergic reaction: once we have read the information leaflet, in the absence of particularly revealing information, we always decide to take it.

7.2.3 Causality, Probabilistic Correlation, and Confirmation of a Scientific Hypothesis

The concept of probability is related to that of cause, although there is a clear-cut difference. Learning reasoning techniques based on this distinction is important for forming an autonomous judgment when a conflict opinion arises among real and self-styled experts. For example, the fact that secondhand smoke makes cancer more likely but does not inevitably cause it may be difficult to understand if we have not learned how to draw a distinction between a probabilistic correlation and causation. A probabilistic correlation between two events is given by the fact that the occurrence of one of the two makes the other more probable (positive correlation) or less probable (negative correlation). For instance, exercise increases the likelihood of enjoying good health but does not inevitably cause it. Heavy smoking makes it more likely for someone to develop cancer but certainly does not cause it. The claim that a heavy smoker whom someone knows lived until 100 is not evidence that smoking is not correlated to cancer but is evidence only for the deficiency of our educational systems, which should have taught us that the frequentist account of probability holds only for very large groups of people and not for individuals. This would be like claiming that a very cold day is sufficient evidence against global warming. Educators unfortunately neglect this wrongheaded kind of reasoning that, however, helps to explain why some people refuse to be cured by treatments that are otherwise statistically very secure.

To understand the difference between probabilistic correlations and causes, consider this example due to the philosopher Wesley C. Salmon (1925–2001). Even if the downward movement of the needle of a barometer makes the arrival of a storm more likely (there is, therefore, a positive probabilistic correlation between the two events), it does not cause it and is not caused by it, since the *common* cause of these two phenomena is a low pressure in the atmosphere. Interestingly, the discovery of common causes has important epistemological consequences.

If it were implausibly discovered that there are genes causing *both* the smoking habits and the onset of cancer, we could not say that smoking is correlated to cancer. Just as in the case of the barometer needle and the arrival of the storm, the probabilistic correlation between the two events (smoking and cancer) would be completely explained by the common cause, which is genetic in nature. It follows that in the absence of this unlikely discovery, we must continue to believe that smoking increases the likelihood of cancer, without causing it.

When we try to establish a correlation, we have to make sure that the sample of elements that we gather is not biased. We tend to select only elements that sustain our beliefs and neglect counterexamples: the origin and reinforcements of all superstitions are explainable in this way. To illustrate, consider the story that sitting thirteen at the table brings bad luck. This is a superstition generated by a passage in the Bible mentioning Jesus sharing a meal with his twelve apostles: as is well known, after the Last Supper, Jesus was betrayed by Judas and then crucified. The number thirteen is not probabilistically correlated to the death of a diner, since if we made many observations, we would find out that the probability that death will occur within a reasonable amount of time following the meal is the *same* as that in which death does not occur. And yet, by selecting and remembering only favorable coincidences, many people who neglect the negative evidence establish their conclusion. The same cognitive errors are in play in astrology and other alternative medical treatments and the mechanism to unmask them is the same.

In our example, to test the superstitious beliefs, we must first specify the interval of time within which the forecast is valid; otherwise, due to its vagueness, the belief would not be empirically controllable. Fifty years after lunch or dinner would be too long, since in that interval of time someone would very probably die. Two days would be too short (it would be unlikely, though not impossible, that someone died in such a short time), while one or two years, despite the arbitrariness of the choice, would be a more reasonable testing interval. Then we would have to collect evidence by using a very large sample of events of the same type: thirteen people at the desk. By organizing the sample in a judicious manner, we would soon discover that it is not the number thirteen that causes death, but other causal factors (health and age) or accidental factor, which we have ignored in trying to prove our original belief.

In a word, the search for causes involves a judicious process of selecting among possible hypotheses. This explains why the notion of cause is important in understanding how a scientific hypothesis is confirmed: the method followed by scientists is analogous to what an investigator does when she proceeds by eliminating one hypothesis after the other.

The following case study, discussed by the philosopher of science Carl Gustav Hempel (1905–97), illustrates how a scientific hypothesis is arrived at and confirmed (1966: chapter 1). In 1844–5, in an obstetrics ward of a Vienna hospital directed by Dr. Ignaz Semmelweiss (1818–65), there were many more deaths due to puerperal fevers than in other wards. Semmelweiss' scientific problem consisted in searching for the cause of this phenomenon and for this purpose he used the so-called method of "eliminative induction."[19] To explain the higher number of deaths in his department, he made various hypotheses (overcrowding, food, etc.) that he rejected one by one by comparing them with what happened in the other wards. After many investigations, he identified the cause of the phenomenon in the fact that before the visit to the department the doctors had been in the morgue for cadaveric dissections. By washing their hands with a suitable disinfectant, the incidence of deaths dropped dramatically, and Semmelweiss' hypothesis was thus decisively confirmed by the statistical data.

This case study shows that even if the scientific disciplines are hyper-branched and require great specialization, a familiarity with "trial-and-error" methods of this type is a good recipe to guide the citizen to realize that there can be different causes explaining a phenomenon. This eliminative method creates the awareness that scientific hypotheses can be false but that the true cause of a phenomenon can be arrived at by eliminating the previous hypotheses. The choice among different positions in public debates involving science or technology can be more rationally taken by becoming aware of this antidogmatic attitude typical of science.

7.2.4 The Role of Consistency in the Evaluation of Two Conflicting Hypotheses

There is another fundamental concept on which the scientific attitude of mind relies, and which can help us decide which of two conflicting theses defended by two groups of experts in a public debate is more plausible. The concept in question is the logical consistency of the beliefs that are brought into play. The "consistency" among a set of beliefs means that there is no belief that contradicts some other ones in the class. In principle, each belief must not contradict any of the others: in epistemology, this position is called "coherentism" (Quine and Ullian 1978). Scientific beliefs (including non-scientific ones) are comparable to nodes of a huge network characterized by mutual logical consistency (see Figure 7.1). A node holds because the threads that bind it to the others are numerous and vice versa. Not all nodes or threads are as robust as others: some beliefs are less solid than others and can be more easily abandoned by cutting the few threads that link them to the other nodes. Some nodes are very robust because they are connected to countless others. The excision of a few disconnected nodes does not destroy the network.

An essential condition for adding one or more nodes to the network is that the new discovered node does not weaken the network, or even break it by cutting off the most robust ones, which correspond to extremely strong, well-confirmed beliefs, contained in the nodes marked in bold character (see figure 7.1 below). Beyond metaphor, the new beliefs that are suggested by self-styled experts and advertised through the web and by other means of mass communication can become part of the reliable or scientific ones only if they do not contradict the countless others that scientists already accept and consider reliable. A belief is founded on science not only because it has been individually tested several times, but especially because it does not contradict many or even all of the other beliefs in the web that have been tested via the complex social procedures illustrated in Chapter 2. If a new hypothesis does not harmonize with many others that are already well established in the social background knowledge, it is unlikely to be true.

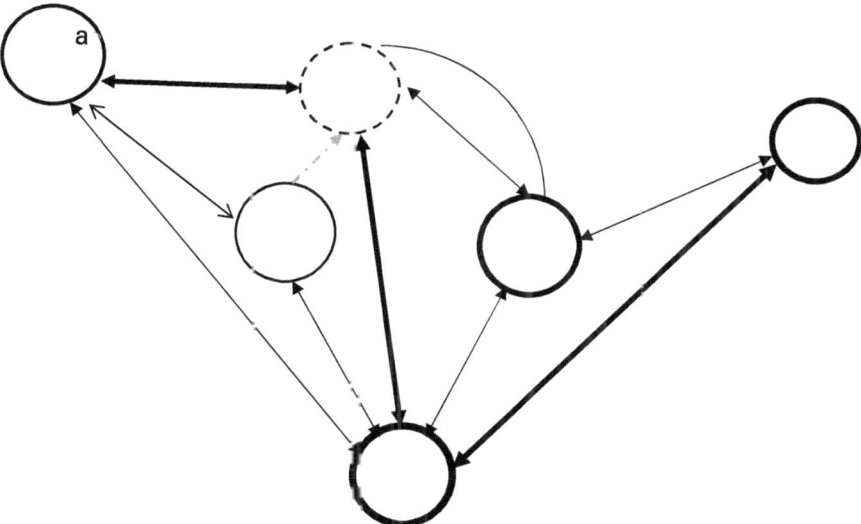

Figure 7.1 Each node represents the beliefs of a community of citizens, or of a group of scientists. The links or arches between the nodes represent the informational exchanges and communication channels between the nodes and therefore the communities. The arches marked in bold represent the logical strength and the coherence of the set of beliefs belonging to two adjacent nodes. The nodes that are marked in bold are formed by more competent communities of experts, while the dashed lines stand for groups of citizens sharing faked news. *Source*: © Mauro Dorato.

One way to express this intuition in a more precise way is in terms of the so-called a priori *probability* of a hypothesis H. For instance, if to accept a new hypothesis H within a network of already accepted ones we had to disavow almost everything that science has taught us so far about the world, the *a priori* probability of the correctness of H would be very low. The "*a priori* probability" is the probability that we should assign to a new belief (the robustness of the new node) in the light of everything scientists regard as established and already know, that is, the already mentioned scientific background knowledge.

This explains why we should accept the thesis that homeopathic treatment, the imposition of hands on the patient's body, or pranotherapy, can have therapeutic effects that can essentially be explained by taking into account the placebo effect (i.e., with the psychological suggestion that is indeed very powerful). If we accepted such "non-standard" beliefs as new knots in the network, we would have to abandon much of the very solid knowledge on which all modern anatomy, physiology, and biochemistry are based. In the case of homeopathy, for instance, we should accept the claim that the active homeopathic principle leaves its traces on water molecules, while in the case of pranotheraphy that there is an exchange of explanatorily relevant energy between the hands of the physician and the patient's mind.

Although we cannot assume that all individuals assign a hypothesis H the same degree of probability (this depends on their competence), the variability in question is harmless if one has a solid background knowledge of science and of its empirical and social basis. Unfortunately, we must observe that there are cases of infringement of the principle of consistency that are even more radical than those underlying homeopathic treatment or pranotherapy.

A Flat Earth society has existed for decades (see: https://theflatearthsociety.org/). The ever-increasing thousands of followers recognizing the principles of this organization believe that the Earth is a flat disk with the North Pole in its center, protected by a 400 km high wall of ice. The flat-earthers, who also organize international conferences (one took place in 2018 in Denver, Colorado) refute with ad hoc hypotheses all the evidence contrary to their beliefs (McIntyre 2021). The ancient Greeks already knew that the Earth is spherical because when the Earth is placed between the Sun and the Moon, it casts a round shadow on the Moon in each eclipse of the Moon, or even at different points of its orbit. If we go up a hill, we see first the mast of the ship and then its keel. The planet has been circumnavigated without encountering ice barriers, planes have circled the world, there are photos of the earth from space, etc. These photos taken from space according to flat-earthers are forged, as are those of the man on the Moon.

The psychological aspect of these bizarre beliefs that should not be overlooked is that, according to flat-earthers, *all the contrary evidence above is the result of a conspiracy*. People should be aware of cases of this kind because the disinformation generated and disseminated by the web can leverage psychological tendencies towards paranoia and the search for the scapegoat that in some of us are more developed than in others. The conspiracy theory of the Masonic Pluto-Jewish democracies was the watchword of fascism, just as the fact that the Jewish financial lobby wanted to dominate the world was the main tool of Hitler's propaganda.[20] Although these criminal beliefs are not even comparable to that of the flat-earthers, in order to fight them we must ask why they spread so effectively.

Science can often sweep away propagandistic. Just to give an example, it is thanks to biology and genetics that we have discovered that the concept of race has no scientific status. If racists had more acquaintance with science, ideological reasons would have had less of a grip and weaker political consequences. Unlike flat-earthers, we believe in the theory of continental drift because the experts in that field, to whose authority we entrust the validity of our beliefs in a vicarious but rational way, base their knowledge on hypotheses supported not only by experiments, and by direct observation (earthquakes, measurable displacement of continents every century, volcanic activity, etc.), but also, indirectly, from everything we already know in other fields of knowledge. Geological beliefs blend harmoniously with physical, chemical, and astronomical beliefs, forming a coherent and homogeneous whole.[21]

This *convergence of hypotheses* is an important element in evaluating the disagreement between experts even in the absence of specialized training and was considered by nineteenth-century British philosopher and scientist William Whewell (1794–1866)[22] the litmus test of scientifically reliable theories (1858). In the history of science, this convergence happened with the belief in the existence of atoms at the beginning of the last century: when it was found that it was possible to calculate the same number (a parameter relating to the existence of atoms)[23] in thirteen different and independent experimental ways, the probability that the belief was false became very low and disagreement among experts (physicists) gave way to consensus.

In sum, These observations on the concepts of probability, correlation, cause, and consistency of beliefs show that an important element allowing us to judge which "expert" we can trust is of a methodological nature, and therefore involves the philosophy of science. This holds to the extent that the latter renders precise concepts having a vague meaning that scientists use implicitly but—understandably from their point of view—without paying attention to their exact conceptual implications.

7.3 The Philosophy of Science and the Refutation of Relativism

In this paragraph, I will reinforce the theses in favor of the objectivity of scientific knowledge already defended at the beginning of Chapter 5. I will refute some arguments in favor of the relativity of the truth of scientific theories (and of truth in general) to different historical periods, to different groups and ethnicities, to people of different sexes, and to non-scientific interests of various kinds.[24] For our purposes, the refutation of this relativist argument is important to reiterate that there are objective facts about reality and that not all interpretations or explanations of such facts are equally valid. Consequently the beliefs of a group of competent individuals do not have the same value as those of another group composed by non-experts. I have said before that many arguments in favor of this equality of weight are based on the claim that both groups are guided by subjective, non-epistemic interests. In this final paragraph, not only will I show that the ideal of the objectivity of knowledge can be achieved but also that the possibility of realizing it has very desirable social consequences.

Relativist theses about human knowledge generally claim that there are no objective facts independent of interpretations or variable criteria of justification. In our context, "interpretations" can be regarded as synonymous with "conflicting hypotheses, defended on the one hand by experts and on the other by self-styled or dishonest experts," with respect to which the public opinion is divided. The relativist affirms that there are no facts that can be invoked to resolve the conflicts of opinions in an objective way and without an interpretation because any apparently neutral description of facts is wholly dependent on practical interests.

If we could justify a relativistic argument about the most solid knowledge we possess (science), we would have to conclude that all our beliefs reflect socially shared points of view that radically change over time and that therefore cannot be considered to be mind-independent descriptions and explanations of facts. However, we have seen why the "incommensurabilist" theses on scientific knowledge—according to a radical version of which Aristotle, Newton, and Einstein did not interpret or explain the same world in a different way but *lived in a different world*,[25]—lend their side to many objections. Even granting that the method of evaluating scientific hypotheses has historically been different across different scientific communities, this would not suffice to justify the thesis that Einstein's theories of spacetime and mass are not more correct than Aristotle's. Or that the most ancient mythological theories on the origin of human beings are on the same level as the Darwinian theories of evolution, by which all contemporary biology is inspired and without which all life sciences would make no sense: "Nothing in Biology Makes Sense Except in the Light of Evolution" (Dobzhansky 1973).

But it is precisely this type of relativism that needs to be argued for in order to show that "one theory is as good as another." If, as the relativists on knowledge affirm, "there are no facts without interpretations," the disagreement between experts could never be resolved on an objective level, because there would be no objectivity coming from observations and other factors involving social procedures for controlling the hypotheses. The prevalence of one theory over another would have an explanation grounded only in non-epistemic values.

According to some relativistic conceptions, this would be another example of a conspiracy theory: the so-called "objective scientific hypotheses" for some relativists are so-classified only to defend the economic interests of large industrial groups financing one research program at the expense of another or filtering certain experimental results from others. This epistemic viewpoint, holding that on the pure epistemic level all hypotheses are equally valid, does not help to eliminate beliefs that are in fact devoid of any scientific status and that are suggested by economic interests or can be reinforced by disinformation traveling on the web. It then happens that these hypotheses are seized by individuals with an inadequate level of scientific culture and moreover without political, social, and economic power, a fact that is certainly not secondary to explain the refusal of "official science." If before a difficult operation, the parents of a 3-year-old child do not want her to receive blood from donors who have been vaccinated against Covid-19 and make a public appeal via chats on Telegram Messenger to find blood from no-vax, something has gone awry.[26]

Another example. If the relativists were right, the placing of a new drug on the market (an indisputable fact F in itself devoid of interest) could be interpreted either as, (i) an attempt at an economic exploitation initiated by a large pharmaceutical company that prescribes its employment (with neutral or harming side-effects) with the help of compliant doctors, or as (ii) a preventive defense of the citizens' health, a thesis supported by other doctors whose opinions conflict with those of the former. The objectivist would argue that the disagreement lie only in the interpretation of the fact F: the drug is either harmful or beneficial in a non-relative way. However, which opinion should we adopt if two groups both apparently composed of authoritative members defend opposite theses because, as the relativist claims, there are just two interpretations and no objective fact?

For the relativist, there is no exclusively logical way to distinguish between the two interpretations, because the facts by themselves "do not speak" and therefore do not support either of the two theses. The appeal to objectivity is purely rhetorical, as it serves to persuade a group of citizens without being logically founded. If there is neither a neutral description of how things are that is objective in the strong sense (to use the terminology introduced in 4.1), nor an objective interpretation in a weak sense, on which all experts or citizens converge, why not leave the choice of the scientific policy to the scrutiny of citizens through a direct vote? It has not been sufficiently noted that also epistemic relativism constitutes a very powerful argument for justifying populism, and direct and non-representative forms of democracy, with the risks that we have already seen.

Suppose with the relativist that—even in those situations in which the evidence would be sufficient to choose one of two policies—the epistemic content of scientific hypotheses involving public health depended completely on the economic and political values of a group of experts, and research communities. Obviously one can only assess on a case-by-case basis whether the interests of a pharmaceutical company prevail over those of citizens, but it is precisely for this reason that relying on the right experts is essential and this can be done at least in part by evaluating the theses defended above.

I will now show why the relativist thesis on facts and epistemic criteria is very implausible by using three logically strong as well as simple arguments. The first is purely pragmatic: the relativist's behavior is not consistent, given that when she needs to be cured of pneumonia, she implicitly relies on scientific theories whose objective credentials in other contexts she denies. This is an implicit trust in science which I described in 4.1 that, however, in other socially very relevant circumstances can be corroded by disinformation.

Since it could be objected that this argument is based only on the practical success of science but does not affect relativist objections, let us move on to discuss the second argument, which is based on *a reductio ad absurdum*:[27] if the relativist thesis on knowledge were correct, there would never be disagreements between experts. But this is precisely the phenomenon that we are trying to focus on, which we often must face and that we must be able to solve by rationally choosing the more probable alternative.[28] Whenever two people (experts included) disagree and engage in a discussion concerning matters of fact (say, whether men have been on the Moon), if they honestly believe in their respective opinions, then they are at the same time *implicitly assuming* that one of them is *right* and the other *wrong*. However, *right and wrong about what*? The notions of "being right" and "being wrong" presuppose that each discussant believes that the fact that she is arguing is true independently of her opinion, or *true simpliciter* (simply true). Otherwise, what would be the point of discussing? The corresponding practice, which characterizes all human societies, would be completely irrational. Suppose relativist A says, "for me, *p* is true," and another relativist B holds at the same time, "for me, *p* is false." If, as the two relativists hold, one cannot find out whether p is true (or false) "absolutely" because p is true (or false) only in relation to each expert's point of view, it follows that A and B cannot disagree. *That is, relatively to both A and B, the claim of the other is correct from the other's point of view*. Since B is a relativist, she accepts the truth of the statement "according to A, *p* is true." Analogously, A accepts the truth of the statement that: "according to B, *p* is false." Paradoxically, there is no disagreement between them about *p*. In other words, the relativist holds both that

P = "neither of the two experts A and B is *right* about what statement *p* expresses in an absolute, non-relational way"

therefore

Q = "neither of them is *wrong* about what statement *p* expresses in an absolute way"

However, this is absurd: if being wrong or right depended on a point of view, the very notion of "being right" in any discussion between individuals with opposite opinions would be meaningless: after endorsing relativism, what would they disagree *about*? The social practice of discussing would be pointless. If there is an agreement between the conflicting parties both on P and Q above, there are *two* other propositions that they both agree upon so that relativism about truth is self-refuting. Relativists cannot, therefore, assume that there is disagreement on *everything*. Similarly, skeptics who affirm that there is no truth ("no assertion or belief is true") should make an exception with regards to the statement expressing their thesis, which is precisely "the only truth is that there is no truth." And this makes their position untenable. These are well-known arguments against epistemic relativism.

The third argument specifies the second by applying it to the relationship between facts and norms. Two dissenting experts must agree on the statement that "the choice of an interpretative hypothesis of a fact otherwise devoid of meaning is always due to conflicting values." But if there is at least one statement (the one in square quotes) on which there is an agreement, then radical relativism is false: at least on the existence of the disagreement due to the different values on which the interpretation of facts depends, there must be agreement also for the relationist. And if there were no agreement on the fact that there is disagreement, there would be agreement on this, too.

The third argument reinforces the anti-relativistic one by applying it to the criteria and epistemic methods for evaluating contrasting hypotheses. That is, suppose that a hypothesis is true for a scientist or a group of experts who uses methodological criteria C_1, and false for another group, which bases its theses on alternative evaluation criteria C_2. Epistemic criteria can include simplicity, the use of the inference to the best explanation, a frequent appeal to falsificationism or inductive methods, etc. If scientist Niels states: "I believe in hypothesis H_1 because I assume criterion C_1," and scientist Albert states: "I believe that H_1 is false because I assume another criterion C_2," both Niels and Albert agree on what they have said to each other. That is, the two statements in the scare quote are true for both, even if they disagree on the criteria to be used to evaluate the other's statements. That is, Niels and Albert both agree on P = "Niels believes the hypothesis H_1 because he presupposes C_1 being valid" and on Q = "Albert believes that H_1 is false because it presupposes another criterion C_2 to be valid."

The third anti-relativist argument is even more lethal because it generates a dilemma between inconsistency and insignificance.

Let us consider the relativist thesis R expressed by the following statement in bold:

R: "Any hypothesis T bold, whether scientific or not, holds with respect to the historical moment M in which it is believed, affirmed, etc." It follows that:

1. either R is true in *all* historical moments

or

2. R is true only in the historical moment **M** in which it is believed, affirmed.

There is no third alternative (tertium non-datur).

In the first case (1) the thesis R is clearly false since there is at least one thesis, precisely R, which does not depend on the historical moment M but is valid for every historical moment, thus becoming a transhistorical, transcontextual, or transparadigmatic truth, something that the relativist cannot accept on pain of self-refutation. This is the first horn of the dilemma.

She must therefore accept the second horn (2), according to which the truth of R depends only on the historical moment **M** in which it is believed. However, at this point, we can raise two further objections.[29]

The first consists in observing that (2) is not strong enough and becomes, as announced above, uninteresting, since it could become false in a historical moment M' other than M. If the relativist is interested in R being true in all historical moments, then we fell into the first—already refuted—case (1) above.

The second objection to (2) consists in saying that if one tries to relativize *also* proposition (2), the relativism that it expresses generates a regress to infinity, which is the second horn of the dilemma. If the truth of T does not hold in all moments but depends only on the historical moment M, one must avoid the charge of inconsistency by trying to relativize also statement (2), thereby obtaining a statement which refers to the truth of R:

R_1 = "R is true only in the historical moment M";
R_1 states:

"R is true only in the historical moment M in which it is believed, affirmed, etc. in the historical moment M

What R_1 states is that every hypothesis (opinion, belief, etc.) expressed by any proposition R holds only in the historical moment M and refers exclusively to M.

However, the same dilemma arises again also in this case. Either thesis R_1 is true in every historical moment and relativism is consequently false, or R_1 is true only in the historical moment M. Boghossian notes that, in this case, we have a meaningless sentence:

> R_i Every truth relates to the historical moment M only to the historical moment M only to historical moment M and so on for each historical moment.

Ultimately, relativism cannot be used to attack the principle of competence on the grounds that no one is more wrong than right. After having refuted the relativistic objection in a way that I believe to be definitive, I must analyze it from another, far more general perspective, which has to do with the possible negative consequences that the objectivity of knowledge may have on the quality of democratic life.

7.4 The Desirability of the Objectivity of Scientific Knowledge

If we admit that scientific knowledge is objective (intersubjectively valid or strongly objective), is not it perhaps true that precisely this objectivity compromises and tends to suppress the pluralism of opinions, of different cultures, and of worldviews that are essential components for any kind of democracy? Should not we conclude that to the extent that science succeeds in reaching the ideal of objective knowledge, it becomes an instrument of oppression of these forms of pluralism?

To answer these important questions, it is enough to expand on some observations already made. Let us assume for a moment that the cognitive objectivity of which science is the supporter is not desirable. To realize the negative consequences of this hypothesis, let us again use the verdicts of a jury and the sentence of a judge as examples, which in Western judicial systems are of fundamental importance to guarantee the freedom of an individual. The jury or the judge can give their sentence in good or bad faith. Let us suppose that the first hypothesis holds: in this case, they may be wrong by unjustly condemning an innocent or acquitting a guilty person.[30]

However, the very possibility of a judicial mistake presupposes that the jury, like all of us, possesses the concept of "correct or fair decision," which has to do with the compliance of the action taken (if done) with the criminal code of the state. "Compliance with a rule," however, presupposes the concept of objective fact to which the action itself corresponds and without which the sentence cannot be triggered. A conviction is unfair if the fact did not happen. So, a decision is "right" or fair because it refers to a fact that either happened or did not happen. If the decisions of a judge were potentially neither *objectively right nor wrong* because the facts on which they are based depended only on the interpretative theory of the accusation or defense, the very notion of justice would lose its meaning. In the presence of overwhelming evidence—think of the recording of a phone call that excludes possible fakes, an objective fact—the jurors in good faith will agree on the same judgment. Why then argue that the intersubjective agreement is an instrument of oppression because it limits the plurality of points of view?

It is obvious that one judge might disagree with another as to whether that particular action requires a sentence of several years or not, but just this more understandable disagreement shows that they both desire that the other should agree with them and therefore that the intersubjective agreement (weak objectivity) based on facts (strong objectivity) would be desirable and constitute a normative presupposition of the practice of dialoguing among human beings.

In support of this position, we can quote a beautiful passage from Popper:

The belief ... in the possibility of legal regulation, equal justice for all, fundamental rights and a free society—can survive the observation that judges are not omniscient, that they can make mistakes about the facts and that, in practice, absolute justice is never fully realized in all granting legal cases. But the belief in the possibility of legal regulation, justice and freedom can hardly survive the acceptance of an epistemology that teaches that there are no objective facts; not only in this case but in any other case; and that the judge cannot have made an error regarding the facts because, regarding them, he cannot be more wrong than he can be right.

1963: 5

Therefore, the desirability of objective knowledge is justified by the fundamental fact that any practice that has to do with justice in a representative democracy *presupposes* scientific objectivity. If the verdict of a trial depended only on human arbitrariness, our social coexistence would be impossible because arbitrariness and the right of the strongest would reign. Of course, it is still possible to reject the value of scientific objectivity and recognize at the same time that science embodies it. After all, The acceptance of a value is a free choice of each of us and objectivity *is* a value By choosing to reject it, one must then provide alternative justifications of all the human practices based on this value that i mentioned before the socially beneficial consequences of endorsing it included itself among them.

Conclusion

In conclusion, I have shown that scientific knowledge is objective in its two meanings (strong and weak), that it progresses thanks to socially established, peer-based criticism of its theories, but ultimately to observations and experiments that eliminate competing hypotheses. Despite the fact that science has a history and is subject to radical changes, cognitive relativism is false. It follows that it is not the case that one theory is as good as another and that all scientific hypotheses with important social repercussions are based only on interests that have nothing to do with the truth. As Plato had it, there is a difference between knowledge and opinion, between competence and ignorance, given that the former, considering our human limitations, typically leads to reliable beliefs, while the latter, driven by disinformation, leads to unreliable guides to action.

To understand on which side to stand and therefore which "experts" are really expert, we need greater scientific literacy. The latter is indispensable for unmasking flat-earthers' type of nonsense and much more dangerous denialist beliefs spreading on the web and threatening the values of a democratic society which, based on the principle of competence, can only be representative. We have also seen why the growth in the level of scientific literacy must be accompanied by a parallel growth in humanistic knowledge, which is connected to the first by the history and philosophy of science. Building a bridge between the two cultures implies the possibility of forming a new generation of citizens equipped with the necessary preparation to orient themselves more consciously in a public debate involving science.

In any case, it is only through knowledge generated by study and culture that respect for the principle of competence—which leads to representative forms of democracies—can be reconciled with the citizens' need to independently determine their goals through their votes and political activity. To conclude with Nelson Mandela, "Education is the most powerful weapon which you can use to change the world" (1990).

Notes

Introduction

1 There are many contemporary philosophers defending this pragmatist attitude. Recalling some of their names would cause injustice to the many others who are not mentioned.
2 The fundamental questions involving the target of scientific literacy and its implications for T_1 will be discussed in depth in sections 6.2 and 6.3.
3 For a good philosophical analysis of the models for climate change, see Frigg, Thompson, and Werndl (2015a-b).
4 See Kourany and Carrier (2020) for the thesis that in the last decades, also scientists have been responsible for the spreading of ignorance.
5 In Chapters 3 and 4, I shall qualify the statement.
6 Kitcher (1990, 2011). Here I regard the adjectives *cognitive* and *epistemic* as synonyms.
7 For the strategy of "raising doubts" in the larger public on the danger of tobacco, see Oreskes and Conway (2010). For a general study of the role of science in producing ignorance, see the already mentioned Kourany and Carrier (2020). The historian of science Proctor (1991) has defined a new branch of knowledge whose main topic is the study of ignorance (agnotology).
8 This expression is taken from the already quoted book of Oreskes and Conway (2010). Also see n. 7.

1 Historical Prologue

1 My thanks to one of the referees for having drawn my attention to this debate.
2 The inverted commas are due to the fact that while Dewey discussed Lippmann's view, there is no trace of replies of Lippmann to Dewey.
3 Even though I will not discuss the secondary literature, the debate is still widely discussed. See, among others, Herbst (1999–2003), Jansen (2008), Westbrook (1991), Brown (2015), and Barrotta (2017: ch. 6, in particular). For recent evaluations of the "debate," see, for instance, Whipple (2005), Herbst (1999–2003), DeCesare (2012), MacGilvray (2010), Feinstein (2015), and Oliverio (2018: ch. 3).

4   *The Phantom Public* is the title of Lippmann's book ([1925] 1993).
5   In Italy, the so-called "Rousseau's platform" officially opened in 2016 by the populist party "Five Stars" attributes a very important role to these kinds of procedures.
6   For more details on this three-partition, see DeCesare's (2012) clear reconstruction of Lippmann ([1922] 1997; and 1925), which I follow here.
7   See Introduction, n. 6.
8   See also the review of Lippmann by Dewey (1927).
9   For Tocqueville, see Chapter 4, section 4.2, in particular.
10 The holistic attitude to knowledge characterizing Dewey's early philosophy has been stressed by Roger's introduction to "The Public and Its Problems," p. 34, which sheds light on the influence that Hegel's holism initially exerted on Dewey's view of knowledge and society.
11 God in Aristotle's language is "pure act," that is, is not subject to any form of becoming or transformation from potentialities to actualities.
12 These are the essential components of Kitcher's view of well-ordered science (2011).
13 Archimedes is a very important exception. For a thought-provoking history of Greek science, see Russo (2004).
14 This is the title of Hacking's (1983) influential book.
15 For his transition to experimentalism, see Dewey ([1929–30] 1984).
16 I am aware that according to Dewey the difference between facts and norms was apparent, and that methodology itself has a normative component.
17 According to Schudson, the positions of Dewey and Lippmann are closer than many scholars have argued: "The intellectual challenge is not to invent a democracy without experts, but to seek a way to harness experts to a legitimately democratic function. In fact, that is exactly what Walter Lippmann intended" (Schudson 2008: 1041). This interpretation is certainly not universally agreed upon by scholars.
18 For the sake of historical accuracy, we should note that this emphasis on a society of knowledge is certainly foreign to Rousseau.
19 For this claim, see below p.60.

2 How Does Science Work?

1   This distinction between descriptions and evaluations is very controversial, especially in pragmatist philosophies, but for my purpose I can postpone its defense in later chapters.

2 When we think of the increasingly powerful tools that science and technology give to human beings, this second thesis is surely true, even though it is important to keep in mind that the *responsibility of the use of these tools is always ours*.
3 The notion of "development or blossoming of human capacities" has been stressed, among others, by the Nobel Prize for Economic Sciences Amartya Sen and by the contemporary philosopher Martha Nussbaum (1998). In the previous chapter, I briefly mentioned these views in presenting Dewey's model of communities.
4 A deductive inference is *valid* when, *if* the premises were true, also the conclusion would be true. A deductive argument is *sound* if and only if it is both valid and all its premises are true. "All dogs are human, Fido is a dog, Fido is human," is an example of a valid deduction that is not sound, since the first premise is false.
5 Today we know that Newtonian mechanics fails whenever bodies travel at a speed close to that of light (where special relativity holds), when the gravitational field is strong (where general relativity applies), and in atomic phenomena (where we need quantum mechanics).
6 There is progress also in mathematics but of a different form. For instance, later proofs of the same theorem are typically much shorter than the preceding ones and generalizations of entire branches of mathematics are not uncommon.
7 For the notion of wavelength, see Figure 2.1.
8 The term *closed society* has been proposed by Popper (1945), that of *static religions* and *morals* has been proposed by Bergson ([1932] 2015).
9 *Learned Ignorance* is the English translation of Nicholas of Cusa's De docta ignorantia [1440] (1990).
10 Voltaire (quoted in Popper 2002: 52). In his publication (2002), Popper has particularly insisted on the relation between epistemology and political philosophy: I have taken these two quotations from his paper.
11 This point has been stressed also by Oreskes (2019).
12 This sentence is attributed to Senator Daniel Patrick Moynihan during a discussion with his electoral opponent in New York.
13 This very low acceptance percentage, however, is explained by considering that, exactly for their importance, these journals receive a number of papers greater than other journals.
14 In this sense, the mechanism of error elimination is a good argument in favor of the fact that, despite some philosopher's reservation (Laudan 1983), there is a distinction or demarcation between science and pseudoscience. For a recent evaluation of the question, see also Pigliucci and Boudry (2013).

15 We refer here to the fact that two adequately prepared systems show a correlation that is not explainable in the classical physics domain and that does not depend on their distance (see Ghirardi, 2015).
16 For what concerns a historical reference to the role of the academies in modern science, see Rossi (2000).
17 Etymologically, in ancient Greek, "cosmopolis" means "city of the world."
18 Among other factors, this negative thesis is based on the very small genetic difference between any two humans, even for geographically separated populations.
19 This attitude is typical of pragmatism and is what Barrotta refers to with the term "scientific mentality" (2017).
20 This definition has been proposed by the sociologist of science Robert K. Merton (1973).
21 In *The Scientific Image*, Bas C. van Fraassen (1980) argues that we should *believe* only in the existence of directly observable components of a theory, but only *accept* as "empirically adequate" its theoretical components.
22 For instance, see Brown (2016) and Skyrms (2004).
23 For an English translation, see Montesquieu (1777).
24 The important reason why "in principle" is italicized will become clear in what follows.
25 This social fact is also characteristic of some religions.
26 Andrei Sakharov, who was a great physicist, fought for the bilateral disarming of the USSR and US anti-ballistic missiles and for the respect of human rights. He was persecuted by the regime and was finally arrested and exiled for his protests against the Soviet invasion of Afghanistan in 1980.

3 How Does Democracy Work?

1 This is a translation of Section 3 of the Italian Constitution, often praised as the most advanced among Western countries.
2 In modern times, the tradition in question (jusnaturalism)) goes back to the jurist and diplomat Hugo Grotius (1583–1645) and the jurist and historian Samuel Pufendorf's (1632–94) and is based on the notion of human nature, where "nature" has a normative aspect. For contemporary translations from Latin, see Grotius (2001) and Pufendorf (2017).
3 In the case of agencies for scientific research programs, the "formal" consists, for instance, in the equality of rules for the submission of projects, anonymous evaluations, etc.

4 In ancient Greek, "demos" is people, while "kratos" is power.
5 See Unger and Smolin (2014: 363–4). Thanks to Giuseppe Trautteur, who called my attention to this quotation.
6 This ideal is mirrored by what Kitcher calls well-ordered science: "science is well-ordered when its specification of the problems to be pursued would be endorsed by an ideal conversation, embodying all human points of view, under conditions of mutual engagement" (2011: 203).
7 In physics, this form of progress holds for Newtonian mechanics, classical electrodynamics, and non-relativistic quantum mechanics.
8 In various European languages, the English word "people" has the same connotations: *Volk, popolo, people*, respectively in German, Italian, and French.
9 This is a recurring theme in Popper's philosophy, at least from his *The Open Society and Its Enemies* (1945).
10 The equivalent Latin expression (*primum non nocere*) is the first Hippocratic rule. Hippocrates (*c.* 460–377 BC) was an ancient Greek physician of the classical age.
11 Unfortunately, the case is different when differences in wealth within a single country and across continents are involved.
12 Here and henceforth, by "expert" I mean what Carrier and Krohn refer to as scientific expert: "Scientific expertise is based on scientific knowledge and responds to demands from society – in particular, politics, economy, and the general public – on how to deal with concrete, typically unprecedented problems" (2018: 55) to distinguish them from technical (carpenters) or professional experts (lawyers).
13 Or at least a partial cause.

4 Representative Democracy, Direct Democracy, and Scientific Specialization

1 Rousseau did not use this term that, however, reflects his political conceptions.
2 It must be noted that in Rousseau's time, suffrage in England was *not* universal. Only in 1928 was the right to vote extended to women.
3 An English translation of his talk is available in Constant (1819). See also Tocqueville (2003).
4 Also, on the possibility to provide incentives to farmers who decide to leave their horns to bovines and ovines!
5 The Nobel Prize winner for Physics in 2021, Giorgio Parisi, has carried out interesting studies on the complex behavior of starlings.

6 In the web, there is a lot of reliable information, but one must know where to look for it.
7 Think of the way in which the *common will* is conceived of, where any difference among individual preferences is weakened.
8 This is not the same question often raised instrumentally (but also very often unattended), based on which a politician can be neither an economist nor an engineer coming from the civil society but can only be elected by the citizens or by other politicians.
9 My translation.
10 I do not want to deny that nutritionists have an important role in our society!
11 Such aims, of course, must comply with ethical principles.
12 The problem related to which experts we should trust represents the object of an extensive debate. The issue has already been addressed in the late 1990s, e.g. see Goldman (2001), Coady (1992), and Kitcher (1993).

5 Scientific Disinformation and the Distrust in Experts

1 Controversies about the most disputed aspects of frontier sciences (involving, for instance, the origin of the universe), which do not have any applicative, social consequences (like global warming or the risks linked to the production of some chemical substances), are not very important for the so-called person in the street.
2 As is well known, together these features define what in one word Kuhn called "paradigm" ([1962] 2012).
3 This division does not imply that these three schools do not share important hypotheses.
4 In the context that is more relevant to the topics I deal with here, a reasonable defense of scientific realism is in Kitcher (2001).
5 For a thorough analysis of the notion of "objectivity in context," see Montuschi (2020).
6 See, for instance, Oreskes and Conway (2010).
7 It must be added that the increasing gap between the rich and poor explains why scientists funded by big companies are often illegitimately regarded as responsible for this lamentable phenomenon.
8 I am a direct witness of this phenomenon since I was a co-editor of an international journal for philosophy of science (EJPS).
9 In the last hundred years, since the physicist Pierre Duhem (1861–1916; see [1907] 1991), historians and philosophers of science have amply investigated this phenomenon.

10 The role of these "epistemic virtues" has been stressed by Kuhn (1977).
11 See, also among others, Weatherall, O'Connor, and Bruner (2020) and O'Connor and Weatherall (2019). In these two paragraphs, I closely follow the narrative of these three texts, especially the first.
12 The methods included funding researchers on the payroll of the industries who brought to light some filtered evidence that was contrary to the much more solid one in favor of the existence of a positive correlation.
13 Among them, Fred Seitz who, after being a scientific counselor of NATO, was elected President of the American National Academy of Sciences first and of the Rockefeller Foundation later.
14 On the basis of rational decision theory, if the probability in favor or P (dangerous consequences of smoking) and not P (smoking is innocuous) are equal, we should do what we desire most.
15 See Barrotta and Montuschi (2018) for the claim that despite the importance of having local sociological factors intervening in the application of empirical knowledge, the latter has an objective character and is valid independently of them.
16 The paternity of this expression is unknown.
17 Disagreement can also be due to the actual desire to go against mainstream theories. As far as climatic change is concerned, for example, it is surprising to find Freeman Dyson (2012), one of the creators of quantum electrodynamics, back up the view that the change is a gigantic mystification.
18 If a test is highly sensitive, the number of false negatives is low. If it is very specific, the number of false positives is low.
19 See Dewey (1910, 1916). For a recent contribution to this problem, see Kitcher (2021b).

6 How to Navigate in the Disagreement of Experts

1 I will not present any data about the extreme poverty of the African continent, of wide areas of the South American continent, and of the Indian subcontinent, which are well known to all.
2 On this point, see Oreskes's criticism (2019).
3 Typical Enlightenment thinker and defender of the thesis of human progress favored by the development of reason, he was a friend of Jean le Rond d'Alembert (1717–83), Voltaire, and the encyclopedists. He participated in the French Revolution, siding against Robespierre, and was eventually imprisoned.
4 The qualification "very probably" reminds us that no belief (or scientific theory) can be certain.

5 Conversely, if the verdict "guilty" is well founded, we assume that the real experts belong to group B: on this hypothesis, the thesis "innocent" defended by group A turns out to be false, because its members are less informed.
6 In slightly more technical language, the probabilities are statistically independent, so that there is some sort of independent control of the correctness of the hypothesis.
7 For simplicity, I consider very probably true (false) as true (false) tout court.
8 Among the most authoritative defenders of this conception of probability, known as "subjectivist," is the mathematician and philosopher Frank Plumpton Ramsey (1903–30) and the Italian statistician Bruno de Finetti (1906–85). For an introduction to the theory of probability and its various interpretations, see Gillies (2000).
9 Assuming that the probabilities in question are identical makes the theorem easier to prove but this assumption is unnecessary.
10 See Anderson (2006) for this suggestion.
11 I thank the anonymous referees for pointing my attention to the importance of the questions that I deal with in this section.
12 With apologies to other authors who have given important contributions to these issues, see, among others, Shen (1975), Feinstein (2010), Phillips, Porticella, Constas, and Bonney (2018), Slater, Huxster, and Bresticker (2019), and Keren (2018). Huxster et. al. (2018) is an extremely important review of the public understanding of science literature in the years 2010–15. For empirical evidence that often the more knowledgeable citizens do not trust science, see Allum, Sturgis, Tabourazi, and Brunton-Smith (2008), Evans and Durant (1995), and Drummond and Fischhoff. (2017).
13 Quoted in Slater, Huxster, and Bresticker (2019: 248), who extensively use this classification.
14 In this respect, see section 4.2 for the important role played by the conceptual difficulties of quantum mechanics.
15 Regarding 6.3.1, I acknowledge important suggestions from one of the referees.
16 Importantly, these authors correctly stress the fact that empirical data may vary from country to county.
17 Vágvölgyi, Dresler, Schrader, and Nuerk (2016).
18 See OECD (2018). Even though the reported data were gathered in 2012, there are no reasons to believe that the situation has significantly improved.
19 Lindemann (2015).
20 For more information about the institution, see OECD, "About PIAAC."
21 Eco (2015).

7 The Role of the History and Philosophy of Science in the Democratic Debate

1. In addition to Rossi (2000), on this point see also Clericuzio (2000).
2. The conflict between the scientific and the manifest has been magisterially described by Sellars (1962), who initiated a long series of studies on this tension.
3. The quotation marks are taken from the title of an essay by Rovelli (2018). The word "always" is mine.
4. The Aristotelian skies move in circular motion.
5. Closer to our solution of the problem, for the philosopher Jean Buridan (circa 1300–58), the stone continues to move because it receives an impetus (a driving virtue) from the hand, which would continue to act forever by making it move in the direction of the throw and with the same speed if weight or friction did not take over. For Buridan, rest and motion still maintain a difference.
6. With respect to the center of the Sun considered to be at rest, the Earth moves along small parts of its orbit in a straight line at approximately 40 km per second.
7. The usual example of a Lamarckian explanation of the length of a giraffe's neck appeals to the effort of the animal to reach the leaves of taller trees *during* its life. The acquired length is then inherited. For my purpose, this slightly simplified treatment will do.
8. This also occurred with creationism, imbued with teleological mental attitudes that have been superseded by evolutionary explanations of the origin of the human being.
9. Here I am referring in particular to the history of last century's physics.
10. Kuhn was talking about losses ([1962] 2012: 99–100).
11. I already referred to quantum mechanics and the obstacle created by the orthodox community to alternative theories and interpretations.
12. For an introduction to epistemology and coherentism, see Audi (2011).
13. Newton's monumental work, published in 1687, was precisely entitled *Philosophia Naturalis Principia Mathematica* (The Mathematical Principles of Natural Philosophy).
14. This metaphor has been proposed by Boghossian (2007).
15. In explications, it is important to reach an ideal equilibrium between two desiderata: on the one hand, very rigorous analysis that completely changed the meaning of the original, intuitive terms would not be faithful to be considered acceptable. On the other hand, without the attempt to make the concept more rigorous, the clarifying attempt would fail.
16. Relevant information was taken from the site Cause.

17 See section 4.4.
18 These data were collected by the Italian network of vaccination systems, published by the Veronesi Foundation's journal.
19 I mentioned this scheme in presenting Popper's philosophy of science, who however denied the existence of induction and described the process referred to by Hempel as a method of conjectures (without any support based on experience) and refutations.
20 A falsified document called *The Protocols of the Elders of Zion* ([1903] 1919) contributed to disseminate hatred toward Jews.
21 In her book, *Why Trust Science?* (2019), Oreskes discusses a case study in which the theory of continental drift was rejected for nationalistic reasons. The important point is that, after some time, consensus was reached.
22 Whewell referred to this convergence as "the consilience of induction."
23 The parameter is the Avogadro number, i.e., the amount of substance that has several particles equal to those of atoms found in 12 grams of carbon 12 atoms (Amedeo Avogadro, 1776–1856).
24 A form of relativism on facts and methodological criteria has been authoritatively defended by the well-known contemporary philosopher Richard Rorty (2003).
25 "As a result of the discovery of oxygen, Lavoisier at least saw nature differently. And if we do not want to somehow make use of the immutable natural hypothesis that he 'saw differently,' the principle of economics will force us to affirm, that after the discovery of oxygen, Lavoisier operated in a different world" (Kuhn [1962] 2012: 147).
26 *La Repubblica* (2022). The minor court had decided in favor of the urgent operation.
27 *A reductio ad absurdum* is an argument in which the assumption of some premises leads to untenable conclusions.
28 In the next paragraphs, I follow closely Boghossian's (2007) magisterial analysis.
29 For this and other anti-relativist arguments, see Nagel (2001) and Boghossian (2007).
30 On the second hypothesis, the jurors know how things are but lie for subjective purposes: it is not a judicial error like the one committed by jurors or judges in good faith.

References

Articles, Chapters, and Books

Allum, N., P. Sturgis, D. Tabourazi, and I. Brunton-Smith (2008), "Science Knowledge and Attitudes across Cultures: A Meta-Analysis," *Public Understanding of Science*, 17 (1): 35–54.

Anderson, E. (2006), "The Epistemology of Democracy," *Episteme: A Journal of Social Epistemology*, 3 (1–2): 8–2.

Audi R. (2011), *Epistemology: A Contemporary Introduction to the Theory of Knowledge*, New York: Routledge.

Bacciagaluppi, G. and A. Valentini (2009), *Quantum Theory at the Crossroads: Reconsidering the 1927 Solvay Conference*, Cambridge: Cambridge University Press.

Barrotta, P. (2017), *Scientists, Democracy and Society: A Community of Inquirers*, Amsterdam: Springer.

Barrotta, P. and E. Montuschi (2018), "Expertise, Relevance and Types of Knowledge," *Social Epistemology*, 32 (6): 387–96. Online at: https://doi.org/10.1080/02691728.2018.1546345.

Bergson, H. (2015), *The Two Sources of Morality and Religion*, Andesite Press.

Bernstein, R. J. (2012), "The Normative Core of the Public Sphere," *Political Theory*, 40 (6): 767–78.

Bobbio, N. (1987), *Future of Democracy*, Cambridge: Polity Press.

Boghossian, P. (2007), *Fear of Knowledge: Against Relativism and Constructivism*, Oxford: Oxford University Press.

Bohman, J. (2010), "Participation through Publics: Did Dewey Answer Lippmann?," *Contemporary Pragmatism*, 7 (1): 49–68.

Brown, J. S. (2016), "Why Darwin Would have Loved Evolutionary Game Theory," *Proceedings: Biological Sciences*, September 14, 283 (1838): 1–9. Online at: 20160847. doi: 10.1098/rspb.2016.0847.

Brown, M. (2013), Book Review, *Science in a Democratic Society*, Minerva, 51: 389–97. Online at: doi.10.1 007/s 1 1024-01 3-9233-y.

Bucci, E. (2015), *Cattivi scienziati. La frode nella ricerca scientifica, prefazione di* Elena Cattaneo, Torino: Add Editore.

Buchler, J. (2014), *The Philosophy of Peirce: Selected Writings*, ed. J. Buchler, New York: Routledge.

Burchell, B. (2016), "Diffraction of Light," Alternative Physics. Online at: http://www.alternativephysics.org/book/Diffraction.htm.

Carnap, R. (1950), *The Logical Foundations of Probability*, Chicago, IL: University of Chicago Press.

Carrier, M. and W. Krohn (2018), "Scientific Expertise: Epistemic and Social Standards—The Example of the German Radiation Protection Commission," *Topoi*, 37: 55–66. Online at: https://doi.org/10.1007/s11245-016-9407-y.

Cartwright, N. (1983), "Causal Laws and Effective Strategies," in *How the Laws of Physics Lie*, 21–43, New York: Oxford University Press.

Cause. "Quote: Wells/Wilks on Statistical Thinking." Online at: https://www.causeweb.org/cause/resources/library/r1266 (accessed August 12, 2022).

Chevassus-au-Louis, N. (2019), *Fraud in the Lab: The High Stakes of Scientific Research*, Cambridge, MA: Harvard University Press.

Christiano, T. (2018), "Democracy," in Edward N. Zalta (ed.), *The Stanford Encyclopedia of Philosophy* (Fall edition). Online at: https://plato.stanford.edu/archives/fall2018/entries/democracy/ (accessed August 6, 2022).

Clagett, M. (1959), *The Science of Mechanics in the Middle Ages*, Madison, WI: University of Wisconsin Press.

Clarke, S. (1717), *A Collection of Papers, which Passed between the Late Learned Mr. Leibniz, and Dr. Clarke, in the Years 1715 and 1716*, London: James Knapton.

Clericuzio, A. (2000), *Elements, Principles and Atoms: Chemistry and Corpuscular Philosophy in the Seventeenth Century*, Dordrecht: Kluwer Academic Publishers.

Coady, C. A. J. (1992), *Testimony: A Philosophical Study*, Oxford: Clarendon Press.

Collins, H. and R. Evans (2017), *Why Democracy Needs Science*, Cambridge: Polity Press.

Condorcet, Marie-Jean-Antoine-Nicolas de Caritat (1785), *Essai sur l'application de l'analyse à la probabilité des décisions rendues à la pluralité des voix*, Paris: Real Academy.

Constant, B. (1819), *Benjamin Constant, The Liberty of Ancients Compared with that of Moderns*, Online Library of Liberty, Liberty Fund. Online at: https://oll.libertyfund.org/page/constant-the-liberty-of-ancients-compared-with-that-of-moderns-1819 (accessed June 29, 2020).

Christiano, T. (2018), "Democracy," in Edward N. Zalta (ed.), *The Stanford Encyclopedia of Philosophy*, Fall edition. Online at: https://plato.stanford.edu/archives/fall2018/entries/democracy/.

Damasio, A. (2006), *Descartes' Error: Emotion, Reason and the Human Brain*, London: Vintage.
Dawkins, R. (1989), *The Selfish Gene*, Oxford; Oxford University Press.
de Regt, H. W. (2017), *Understanding Scientific Understanding*, Cambridge: Cambridge University Press.
DeCesare, T. (2012), "The Lippmann-Dewey 'Debate' Revisited: The Problem of Knowledge and the Role of Experts in Modern Democratic Theory," *Philosophical Studies in Education*, 43: 106–15.
Dewey, J. (1888), *The Ethics of Democracy*, University of Michigan, Philosophical Papers, Second Series, No. 1, Ann Arbor, MI: Andrews & Company Publishers, 227.
Dewey, J. ([1907–9] 1977), "The Influence of Darwinism on Philosophy," in J.A. Boydston (ed.), *The Middle Works of John Dewey*, vol. 4: *1907–1909*, 7–8, Carbondale and Edwardsville, IL: Southern Illinois University Press.
Dewey, J. (1910), *How We Think*, Chicago, IL: Gateway Edition.
Dewey, J. (1916), Democracy and Education: An Introduction to the Philosophy of Education, New York: Macmillan.
Dewey, J. (1922), "Public Opinion," *The New Republic*, 30 (May 3): 286–8.
Dewey, J. ([1922] 1976), *The Middle Works of John Dewey, 1899-1924*, Carbondale, IL: Southern Illinois University Press.
Dewey, J. (1925a), *Experience and Nature*, Chicago, IL: Open Court Publishing.
Dewey, J. (1925b), "Practical Democracy," *The New Republic*, 45 (December 2): 52–4.
Dewey, J. (1927), *The Public and Its Problems*, New York: Henry Holt.
Dewey, J. ([1929–30] 1984), "From Absolutism to Experimentalism," in J. A. Boydston (ed.), *The Later Works of John Dewey*, vol. 5, 156, Carbondale and Edwardsville, IL: Southern Illinois University Press.
Dobzhansky, T. (1973), "Nothing in Biology Makes Sense Except in the Light of Evolution," *American Biology Teacher*, 35 (3): 125–9.
Dorato, M. (2004), "Epistemic and Non-epistemic Values in Science", in P. K. Machamer and G. Wolters (eds.), *Science, Values, and Objectivity*, 52–77, Pittsburgh, PA: University of Pittsburgh Press.
Dorato, M. (2011), "How to Combine and Not to Combine Physics and Metaphysics," in V. Karakostas and D. Dieks (eds.), *EPSA11 Perspectives and Foundational Problems in Philosophy of Science*, The European Philosophy of Science Association Proceedings 2, 295–305, Cham: Springer International Publishing. Online at: DOI 10.1007/978-3-319-01306-0_24.
Douglas, H. (2000), "Inductive Risks and Values in Science," *Philosophy of Science*, 67: 559–79.

Douglas, H. (2009), *Science, Policy, and the Value-Free Ideal*, Pittsburgh, PA: University of Pittsburgh Press.

Drummond, C. and B. Fischhoff (2017), "Individuals with Greater Science Literacy and Education Have More Polarized Beliefs on Controversial Science Topics," *Proceedings of the National Academy of Sciences*, 114 (36): 9587–92.

Duhem, P. ([1907] 1991), *The Aim and Structure of Physical Theory*, Princeton, NJ: Princeton University Press.

Durkheim, E. (1951), *Suicide: A Study in Sociology*, trans. J. A. Spaulding and G. Simpson, ed. with introduction by G. Simpson, New York: The Free Press.

Dyson, F. (2012), "Freeman Dyson: Climate Change Predictions are Absurd." Online at: https://www.youtube.com/watch?v=fmy0tXcNTPs (accessed August 6, 2022).

Eco, U. (2015), *La Stampa*, June 11.

Einstein, A. ([1933] 1954), "On the Method of Theoretical Physics," in A. Einstein, *Ideas and Opinions*, 270, New York: Crown Publish Einstein.

Einstein, A. (1954), *Ideas and Opinions*, New York: Crown Publish Einstein.

Einstein, A. (1979), *Autobiografia scientifica*, Torino: Universale scientifica Boringhieri.

Einstein, A. (1986), *The Collected Papers of Albert Einstein*, Princeton, NJ: Princeton University Press.

Einstein, A., M. Born and H. Born (2004), *The Born–Einstein Letters 1916–55: Friendship, Politics and Physics in Uncertain Times*, New York: Palgrave Macmillan.

Evans, G. and J. Durant (1995), "The Relationship between Knowledge and Attitudes in the Public Understanding of Science in Britain," *Public Understanding of Science*, 4 (1): 57–74.

Feinstein, N. (2015), "Education, Communication, and Science," *Journal of Research in Science Teaching*, 52 (2): 145–63.

Fernstein, N. (2010), "Salvaging Science Literacy," *Issues and Trends*, Wiley Online Library, 169–83. Online at: Doi 10.1002/sce.20414, wileyonlinelibrary.com.

Feyerabend, P. (1975), *Against Method: Outline of an Anarchistic Theory of Knowledge*, London: New Left Books.

Feyerabend, P. (1978), *Science in a Free Society*, London: Verso Editions/NLB (New Left Books).

Fishkin, J. (2009), *When the People Speak*, New York: Oxford University Press.

Floridi, L. (2014), "A Defence of Constructionism: Philosophy as Conceptual Engineering," in Wiley Online Library, April 4. Online at: https://onlinelibrary.wiley.com/doi/abs/10.1111/j.1467-9973.2011.01693.x (accessed August 9, 2022).

Frigg, R., E. Thompson, and C. Werndl (2015a), "Philosophy of Climate Science Part I: Observing Climate Change," *Philosophy Compass*, 10 (12): 953–64.

Frigg, R., E. Thompson, and C. Werndl (2015b), "Philosophy of Climate Science Part I: Modelling Climate Change," *Philosophy Compass*, 10 (12): 965–77.

Galison, P. (1987), *How Experiments End*, Chicago, IL: University of Chicago Press.

Galison, P. (1987), *How Experiments End*, University of Chicago Press, Chicago.

Galison, P. and B. Hevly, eds. (1992), *Big Science: The Growth of Large-Scale Research*, Stanford, CA: Stanford University Press.

Ghirardi, G. C. (2005), *Sneaking a Look at God's Cards: Unraveling the Mysteries of Quantum Mechanics*, Princeton, NJ: Princeton University Press.

Giere, R. (1988), *Explaining Science*, Chicago, IL: University of Chicago Press.

Gigerenzer, G. (2007), *Gut Feelings: Short Cuts to Better Decision Making*, London: Viking.

Gillies, D. (2000), *Philosophical Theories of Probability*, London: Routledge.

Glymour, C. (1989), *Thinking Things Through*, Cambridge, MA: The MIT Press.

Goldman, A. (2001), "Which Experts Should We Trust?," *Philosophy and Phenomenological Research*, 43 (1): 85–110.

Grotius, H. (2001), *On the Law of War and Peace*, Kitchener, Ontario: Batoche Book.

Hacking, I. (1983), *Representing and Intervening*, Cambridge: Cambridge University Press.

Hempel, C. G. (1965), *Aspects of Scientific Explanation*, New York: The Free Press.

Hempel, C. G. (1966), *Philosophy of Natural Science*, Upper Saddle River, NJ: Prentice-Hall.

Herbst, S. (1999–2003), "Lippman's Public Opinion Revisited," *The MIT Press: Harvard International Journal of Press/Politics*, 4 (2): 88–9.

Hobbes, T. (2012), *Leviathan*, ed. Noel Malcolm, 3 vols, Oxford: Clarendon Press.

Howard, D. A. (2017), "Einstein's Philosophy of Science," in Edward N. Zalta (ed.), *The Stanford Encyclopedia of Philosophy*, Fall edition. Online at: https://plato.stanford.edu/archives/fall2017/entries/einstein-philscience/ (accessed August 6, 2022).

Huff, D. (1954), *How to Lie with Statistics*, New York: W. W. Norton.

Huggett, N. (2010), *Everywhere and Everywhen: Adventures in Physics and Philosophy*, Oxford: Oxford University Press.

Hull, D. L. (1988), *Science as a Process: An Evolutionary Account of the Social and Conceptual Development of Science*, Chicago, IL: Chicago University Press.

Hume, D. (2000), *A Treatise of Human Nature*, Oxford: Oxford University Press.

Huxster, J. K., M. H. Slater, J. Leddington, V. LoPiccolo, J. Bergman, M. Jones, C. McGlynn, N. Diaz, N. Aspinall, J. Bresticker, and M. Hopkins (2018), "Understanding 'Understanding,'" *Public Understanding of Science*, 27 (7): 756–71.

Jansen, S. C. (2008), "Walter Lippmann, Straw Man of Communication Research," in D. D. Park and J. Pooley (eds.), *History of Media and Communication Research*, 71–112, New York: Contested Memories.

Kant I. (1784), "Beantwortung der Frage: Was ist Aufklärung?," *Berlinische Monatsschrift* (December- Heft 1784), S. 481–494.

Kant, I. ([1793] 2009), *Religion within the Bounds of Bare Reason*, Indianapolis, IN: Hackett Publishing.

Keren, A. (2018), "The Public Understanding of What? Laypersons' Epistemic Needs, the Division of Cognitive Labor, and the Demarcation of Science," *Philosophy of Science*, 85 (5): 781–92.

Kitcher, P. (1984), *The Nature of Mathematical Knowledge*, New York: Oxford University Press.

Kitcher, P. (1990), "The Division of Cognitive Labor," *Journal of Philosophy*, 87 (1): 5–22.

Kitcher, P. (1993), *The Advancement of Science*, New York: Oxford University Press.

Kitcher, P. (2001), *Science, Truth, and Democracy*, New York: Oxford University Press.

Kitcher, P. (2010), "The Division of Cognitive Labor," *Journal of Philosophy*, 87 (1): 5–22 (1990).

Kitcher, P. (2011), *Science in a Democratic Society*, New York: Prometheus Books.

Kitcher, P. (2012), *Preludes to Pragmatism: Toward a Reconstruction of Philosophy*, New York: Oxford University Press.

Kitcher, P. (2021a), *Moral Progress*, with A. Srinivasan, S. Neiman, and R. Jeaggi, ed. J.-C. Heilinger, New York: Oxford University Press.

Kitcher, P. (2021b), *The Main Enterprise of the World: Rethinking Education*, New York: Oxford University Press.

Kohn, A. (1987), *False Prophets: Fraud, Error, and Misdemeanor in Science and Medicine*, New Yoork: Blackwell Publishers.

Kourany, J. and Carrier, M., eds. (2020), *Science and the Production of Ignorance: When the Quest for Knowledge is Thwarted*, Cambridge, MA: The MIT Press.

Koyré, A. (1965), *Newtonian Studies*, Cambridge, MA: Harvard University Press.

Kuhn, T. ([1962] 2012), *The Structure of Scientific Revolutions*, Chicago, IL: University of Chicago Press.

Kuhn, T. (1977), *The Essential Tension: Selected Studies in Scientific Tradition and Change*, Chicago, IL: University of Chicago Press.

La Repubblica (2022), February 8.

Laudan, L. (1983), "The Demise of the Demarcation Problem," in R. S. Cohen and L. Laudan (eds.), *Physics, Philosophy and Psychoanalysis: Essays in Honor of Adolf Grünbaum*, 111–27, Dordrecht: Reidel.

Lewis, D. (1980), "A Subjectivist Guide to Objective Chance," in R. Carnap and R. Jeffrey (eds.), *Studies in Inductive Logic and Probability*, 263–93, Berkeley, CA: University of California Press.

Lindemann, R. (2015), "(Il)literacy in the European Union," TextAid, August 20. Online at: https://textaid.readspeaker.com/illiteracy-in-the-european-union/ (accessed August 9, 2022).

Lippmann, W. ([1922] 1997), *Public Opinion*, reprint, New York: The Free Press.

Lippmann, W. ([1925] 1993), *The Phantom Public*, reprint, New York: Harcourt Brace.

Locke, J. (2006), *An Essay Concerning Toleration: And Other Writings on Law and Politics, 1667–1683*, ed. J. R. Milton and Philip Milton, New York: Oxford University Press.

MacGilvray, E. (2010), "Dewey's Public," *Contemporary Pragmatism*, 7 (1): 31–47.

Mandela, N. (1990), Reference Speech, Madison Park High School, Boston, MA, 23 June.

McCloskey, M., A. Caramazza, and B. Gree (1980), "Curvilinear Motion in the Absence of External Forces: Naive Beliefs about the Motion of Objects," *Science*, 210: 1139–41.

Merton, R. K. (1973), *The Sociology of Science: Theoretical and Empirical Investigations*, Chicago, IL: University of Chicago Press.

Mill, J. S. ([1859] 2015), *On Liberty, Utilitarianism and Other Essays*, ed. with introduction and notes by M. Philp and F. Rosen, Oxford: Oxford University Press.

McIntyre, L. (2021), *How to Talk to a Science Denier: Conversations with Flat Earthers, Climate Deniers, and Others Who Defy Reason*, Cambridge, MA: The MIT Press.

Montesquieu, C. L. (1777), *The Spirit of the Laws: The Complete Works of M. de Montesquieu*, vol. 1, London: T. Evans.

Montesquieu, C. (2015), *Lo spirito delle leggi*, Torino: Utet.

Montuschi, E. (2020), "Finding a Context for Objectivity," *Synthese*, 199 (2021): 4061–76. Online at: https://doi.org/10.1007/s11229-020-02969-6.

Morris, D. (1967), *The Naked Ape. A Zoologist's Study of the Human Animal*, New York: Random House.

More, T. (2003), *Utopia*, London: Penguin Classics.

Nagel, T. (2001), *The Last World*, New York: Oxford University Press.

NewsGuard (2020), "Real-Time Reporting on COVID-19 Misinformation." Online at: https://www.newsguardtech.com/covid-19-resources/ (accessed August 6, 2022).

Newton, I. (1998), *Opticks: Or, a Treatise of the Reflexions, Refractions, Inflexions and Colours of Light, Also Two Treatises of the Species and Magnitude of Curvilinear Figures*, Commentary by Nicholas Humez, Palo Alto, CA: Octavo.

Newton, I. ([1726] 2016), *The Principia: The Authoritative Translation and Guide: Mathematical Principles of Natural Philosophy*, Berkeley and Los Angeles, CA: University of California Press.

Nicholas of Cusa (1990), *Nicholas of Cusa on Learned Ignorance*, Minneapolis, MN: The Arthur J. Banning Press. Full text accessible online at: https://www.jasper-hopkins.info/DI-Intro12-2000.pdf (accessed August 10, 2022).

Norr, R. (1952), "Cancer by the Carton: Recent Medical Researches on the Relationship of Smoking and Lung Cancer," condensed from *Christian Herald*, *Reader's Digest* (December): 7–8.

Norris, S. P. and L. M. Phillips (2003), "How Literacy in Its Fundamental Sense is Central to Scientific Literacy," *Science Education*, 87 (2): 224–40.

Nozick, R. (2001), *Invariances: The Structure of the Objective World*, Cambridge, MA: Belknap Press.

Nussbaum, M. (1998), *Cultivating Humanity: A Classical Defense of Reform in Liberal Education*, Cambridge, MA: Harvard University Press.

Nussbaum, M. (2011), *Creating Capabilities. The Human Development Approach*, Cambridge, MA: Belknap Press.

O'Connor, C. and J. O. Weatherall (2019), *The Misinformation Age*, New Haven, CT, and London: Yale University Press.

Oliverio, S. (2018), *La filosofia dell'educazione come "termine medio,"* Lecce: Pensa.

Oreskes, N. (2019), *Why Trust Science?*, Princeton, NJ: Princeton University Press.

Oreskes, N. and E. M. Conway (2010), *Merchants of Doubt*, New York: Bloomsbury Publishing.

Organisation for Economic Co-operation and Development (OECD). "About PIAAC." Online at: https://www.oecd.org/skills/piaac/ (accessed August 9, 2022).

(OECD (2018), Programme for International Student Assessment (PISA). Online at: https://www.oecd.org/pisa/#:~:text=PISA%20is%20the%20OECD's%20 Programme,to%20meet%20real%2 (accessed August 6, 2022).

Pariser, E. (2012), *The Filter Bubble: How the New Personalized Web is Changing What We Read and How We Think*, London: Penguin.

Phillips, T., N. Porticella, M. Constas, and R. Bonney (2018), "A Framework for Articulating and Measuring Individual Learning Outcomes from Participation in Citizen Science," *Citizen Science: Theory and Practice*, 3 (2): 1–19. Online at: Doi.org/10.5334/cstp.126.

Pigliucci, M. and M. Boudry, eds. (2013), *Philosophy of Pseudoscience: Reconsidering the Demarcation*, Chicago, IL: University of Chicago Press.

Poincaré, H. ([1902] 2017), *Science and Hypothesis*, New York: Bloomsbury Academic.

Popper, K. R. (1945), *The Open Society and Its Enemies*, 2 vols, London: Routledge.

Popper, K. (1957), *The Poverty of Historicism*, London: Routledge.

Popper, K. R. (1963), *Conjecture and Refutations*, London: Routledge and Keegan Paul.

Popper K. (1972), *Objective Knowledge: An Evolutionary Approach*, Oxford: Oxford University Press.

Popper, K. (2002), *Conjectures and Refutations: The Growth of Scientific Knowledge*, London: Routledge.
Proctor, R. (1991), *Value-Free Science?*, Cambridge, MA: Harvard University Press.
Pufendorf, S. (2017), *Of the Law of Nature and Nations: In Five Volumes*, Indianapolis, IN: Liberty Fund.
Putnam, H. (2004), *The Collapse of the Fact Value Distinction and Other Essays*, Cambridge, MA: Harvard University Press.
Quine, W. V. and J. S. Ullan (1978), *The Web of Belief*, New York: McGraw-Hill.
Rawls, J. (1971), *A Theory of Justice*, Cambridge, MA: Harvard University Press.
Rorty, R. (1998), *Truth and Progress: Philosophical Papers*, New York: Cambridge University Press.
Rorty, R. (1979), *Philosophy and the Mirror of Nature*, Princeton: Princeton University Press.
Rossi, P. (2000), *The Birth of Modern Science*, Oxford: Blackwell Publishers.
Rousseau, J.-J. ([1762] 1997), *The Social Contract and Other Later Political Writings*, Victor Gourevitch (ed. and trans.), Cambridge: Cambridge University Press.
Rovelli, C. (2018), *The Order of Time*, New York: Riverhead Books.
Russell, B. (1917), *Mysticism and Logic and Other Essays*, London: G. Allen & Unwin.
Russo, L. (2004), The Forgotten Revolution: How Science Was Born in 300 BC and Why It Had to be Reborn, Cham: Springer.
Schiffrin, A. (2017), "How Europe Fights Fake News," *Columbia Journalism Review*, October 26. Online at: https://www.cjr.org/watchdog/europe-fights-fake-news-facebook-twitter-google.php (accessed August 6, 2022).
Schudson, M. (2008), "The 'Lippmann–Dewey Debate' and the Invention of Walter Lippmann as an Anti-Democrat 1986–1996," *International Journal of Communication*, 2: 1031–42.
Scopus. Online at: https://www.scopus.com/search/form.uri?display=authorLookup#author (accessed August 9, 2022).
Sellars, W. (1962), "Philosophy and the Scientific Image of Man," in R. Colodny (ed.), *Frontiers of Science and Philosophy*, 35–78, Pittsburgh, PA: University of Pittsburgh Press; reprinted in *Science, Perception and Reality*, London: Routledge & Kegan Paul, and New York: The Humanities Press, 1963, pp. 1–40; and reissued Atascadero, CA: Ridgeview Publishing, 1991; and in K. Scharp and R. B. Brandom (eds.), *In the Space of Reasons: Selected Essays of Wilfrid Sellars*, 369–408, Cambridge, MA: Harvard University Press, 2007.
Sen, A. (2000), *Development as Freedom*, New York: Anchor Books
Shamos, M. H. (1995), The Myth of Scientific Literacy, New Brunswick, NJ: Rutgers University Press.
Shen, B. S. P. (1975), "Science Literacy," *American Scientist*, 63 (3): 265–8.

Sidereus Nuncius (1610), Wikipedia. Online at: https://it.wikipedia.org/wiki/Sidereus_Nuncius (accessed August 6, 2022).

Silvestri, M. (1978–82), *La decadenza dell'Europa occidentale*, 4 vols, Torino: Einaudi.

Simon, H. (1983), *Models of Bounded Rationality*, 2 vols, Cambridge, MA: Cambridge University Press.

Skyrms, B. (2004), *The Stag Hunt and the Evolution of the Social Contract*, Cambridge: Cambridge University Press.

Slater, M. H., J. K. Huxster, and J. E. Bresticker (2019), "Understanding and Trusting Science," *Journal for General Philosophy of Science*, 50: 247–61. Online at: doi.org/10.1007/s10838-019-09447-9.

Sober, E. and D. S. Wilson (1998), *Unto Others: The Evolution and Psychology of Unselfish Behavior*, Cambridge, MA: Harvard University Press.

Snow, C. P. ([1959] 2013), *The Two Cultures and the Scientific Revolutions*, Mansfield Center, CT: Martino Fine Books.

Temming, M. (2018a), "Computer Programs Call Out Lies on the Internet," *Science News*, 4 August, 194 (3): 22–6.

Temming, M. (2018b), "On Twitter, the Lure of Fake News is Stronger than the Truth," *Science News*, 3 March, 193 (6): 14.

Tocqueville, A. ([1831] 2003), *Democracy in America and Two Essays on America*, London: Penguin Classics.

Unger, R. M. and L. Smolin (2014), *The Singular Universe and the Reality of Time: A Proposal in Natural Philosophy*, Cambridge: Cambridge University Press.

United Nations Educational, Scientific and Cultural Organization (UNESCO). Online at: https://unevoc.unesco.org/home/TVETipedia+Glossary/filt=all/id=708 (accessed August 9, 2022).

Vágvölgyi, R., A. Coldea, T. Dresler, J. Schrader, and H.-C. Nuerk (2016), "A Review about Functional Illiteracy: Definition, Cognitive, Linguistic, and Numerical Aspects," *Frontiers in Psychology*, 7: 1617. Online at: doi.org/10.3389/fpsyg.2016.01617.

van Fraassen, Bas C. (1980), *The Scientific Image*, Oxford: Oxford University Press.

Voltaire ([1763] 2000), *Treatise on Tolerance*, Cambridge: Cambridge University Press. Online at: https://earlymoderntexts.com/assets/pdfs/voltaire1763.pdf (accessed August 6, 2022).

Weatherall, J. O., C. O'Connor, and J. P. Bruner (2020), "How to Beat Science and Influence People: Policy Makers and Propaganda in Epistemic Networks," *British Journal for the Philosophy of Science*, 71 (4): 1157–86.

Web of Science. Online at: www.webofknowledge.com (accessed August 9, 2022).

Weber, M. ([1904–5] 1992), *The Protestant Ethic and the Spirit of Capitalism*, trans. T. Parsons, introduction by A. Giddens, London: Routledge.

Weber, M. (1949), "'Objectivity' in Social Science and Social Policy," in E. A. Shils and H. A. Finch (trans. and ed.), *Methodology of the Social Sciences*, 50–112, Glencoe, IL: The Free Press.

Weber, M. (1958), "Science as a Vocation," in H. Gerth and C. W. Mills (eds.), *From Max Weber*, 125–56, New York: Oxford University Press.

Westbrook, R. (1991), *John Dewey and American Democracy*, New York: Cornell University Press.

Whewell, W. (1858), *Novum Organon Renovatum*, London: J. W. Parker and Son.

Whipple, M. (2005), "The Dewey–Lippmann Debate Today: Communication Distortions, Reflective Agency, and Participatory Democracy," *Sociological Theory*, 23 (2): 156–78.

Wikimedia. Online at: https://upload.wikimedia.org/wikipedia/commons/thumb/c/cf/EM_Spectrum_Properties_edit.svg/2560px-EM_Spectrum_Properties_edit.svg.png (accessed August 6, 2022).

Wilks, S. S. (1951), "Presidential Address to the American Statistical Association," *Journal of American Statistical Association*, 46 (253): 1–18.

Index

academies of science 38–9
agents and spectators 13
Allum, N. 116
American Declaration of Independence 47, 48
ancient Greece 17, 34–5, 64, 66–7
Anderson, E. 111
Aristotle 17, 128, 129
autonomous choice 64–5, 70–3, 73–5, 93–4, 102, 108

bad luck 140–1
Bobbio, Norberto 58
Bohr, Niels 38
Bucci, Enrico 97–8

causal relationships 59–60, 139–41
change 30–3, 130–4
charlatans and swindlers 88–92
citizens
  benefits of science for 39
  decision-making by 51, 64–7, 69
  gap between experts and 12–16, 19–23
  in a technocracy 45–6
  see also autonomous choice; equality of rights
citizen's committees/councils 14
Cline, Martin J. 93–4, 95
common sense 77, 127–30
common will 65–6
communication between scientists 39
  see also publication of scientific works
communities
  democracies and 47, 65
  ideal 19–20
  modern 20–1
  scientific 33–4, 36, 42, 53, 56–7, 76, 133
competence
  evaluation of 121–4
  principle of 107

of representative democracies 2, 70, 71, 77
competent outsiders 113, 114–15
Condorcet, Nicolas de 106–13
consensus
  political 43, 54, 57–8
  scientific 3, 34, 36–7, 40, 42, 52, 83–4, 86
consistency, principle of 133, 142–5
conspiracy theories 144–5, 147
Constant, Benjamin 66, 67
control of science 35–40, 56, 76–8
convergence of hypotheses 145
Conway, E. M. 93–4
cooperation 41–2, 60
crime 119
criticism in a democracy 42–5, 54
cultural gap 120

databases 122
Dawkins, R. 41
deaths in hospital 141
debates, public 89
DeCesare, T. 13, 15
decision-makers 13
decision-making
  by citizens 51, 64–7, 69
  by direct democracy 57, 103
Declaration of American Independence 47, 48
deductive inferences 28–9
democracy 3–4, 4–5, 47–61, 63–79
  autonomous choice 73–5
  characteristics of 63
  control of science 76–8
  criticism in 42–5, 54
  direct democracy and populism 64–9
  disagreement among experts 78–9
  equality principle 47–51
  forms of 11
  majority principle 51–5, 106, 109, 112

pluralism, benefit of 60–1
problem-solving, as an aim 58–60
separation of powers principle 55–8
technocracy, risks of 69–73
Dewey, John 10, 14, 16–18
Dewey–Lippmann debate 2–3, 9–23
   Dewey's criticism of Lippmann 19–23
   gap between experts and citizens 12–16
   knowledge, transmission and distribution of 12–13
   methodological component of Dewey's criticism 16–18
   models of democracy 11–12, 19
   reconciliation problem 9–10
Di Bella, Luigi 90
dictatorial regimes 44
direct democracy 5
   arguments against 66–7
   autonomous choice and 64–5
   characteristics of 63
   comparison with representative democracy 11–12, 64, 65
   decision-making by 57, 103
   as model of democracy 11
   pluralism in 63
   and populism 64–9
disinformation and distrust 2, 6, 81–103
   disagreement among experts 88–97
   doubts and trust 92–7
   evidence, lack and uses of 91–2
   fraud and inductive risk 97–102
   scientific ethics 92–7
   swindlers and charlatans 88–91
   trust in science 82–7
   see also trust
distributive justice 50
Douglas, H. 100, 101–2
Durant, J. 116

Eco, Umberto 121
education 10, 14–15, 16, 22
   see also illiteracy; scientific literacy, need for
Einstein, Albert 16, 32–3, 38, 53, 91, 125–6
electromagnetic radiation 27, *27*
epistemic distances 15–16
epistemic values 96–7, 99–100, 102
equality before the law 48

equality of rights 43–4, 48–9
equality principle in democracy 47–51
ethical stances 117
ethics, scientific 92–7
Euclidean geometry 30
Evans, G. 116
evolution of ideas 40–2
evolutionary theory 18
experiments 17–18, 71–2
experts
   disagreement among 78–9
   and free choice 73–5
   gap between citizens and 12–16, 19–23
   politicians as 71
   relation with politicians 13, 15, 21, 45–6
   role of 13, 15
   trust in 45, 74–5
explications 137

facts, independence of 37
fake news 137–9
fallibility of scientific knowledge 33–5
Feinstein, N. 113
Feyerabend, Paul 131
first thesis of the book *see* specialization of scientific knowledge
flat-earthers 144
fraud, scientific 97–8
freedom of individuals 66, 73–5
functional illiteracy 117–21

gap, cultural 120
general theory of relativity *32*, 32–3, 132
Greece, ancient 17, 34–5, 64, 66–7

h-index 122
Hacking, I. 18
harm, prevention of 56
health *see* smoking and health
Hempel, Carl Gustav 141
history and philosophy, role of 6–7, 125–54
   common sense, overcoming of 127–30
   history of science 126–34
   objectivity, desirability of 152–4
   philosophy, roles of 134–45
   refutation of relativism 146–52

scientific change and acceptance of theories 30–3, 130–4
Hitler, Adolf 51
Huff, Darren 94
Huxster, J. K. 114

ideal types 11
ideas, evolution of 40–2
ideological distortion of science 39, 116
ignorance 2, 34–5, 101
illiteracy 115–21
    see also scientific literacy, need for
inductive inferences 29–30
inductive risk 99, 100, 101, 102
inequality of rights 44
inequality of scientists 68
inertia, law of 128–9
inferences in science 28–30
internet
    evaluation of competence 121–4
    fake news 118
    false theories spread on 102
    science and 67–9
    use in democracies 67
intuition 110
Italian Constitution 49

journals see publication of scientific works
judiciary power 57
jury theorem 6, 106–13
justice 50
justifiability 44–5

Kant, I. 72–3
Keren, A. 112
Keynes, John M. 137–8
Kitcher, P. 14, 23
knowledge, theory of 16–17
Kuhn, Thomas 68, 83, 132

learned ignorance 34–5
legislative power 56–7
letters, of scientists 39
light, theories of 30–1, 31, 53
Lippmann, Walter 12–13, 14–15, 16
    see also Dewey–Lippmann debate
literacy see illiteracy; scientific literacy, need for

logical consistency see consistency, principle of
logical questions 52
luck 140–1

magical healing 90
Maji Maji Rebellion 90
majority principle 51–5, 106, 109, 112
majority, tyranny of 55
Mangabeira, Roberto 53
media, debates held by 89
medical testing 99–100
memes 41
Merton, R. K 96
Mill, John Stuart 34, 55
minority, contribution of 53, 54, 55, 56
models of democracy 11–12, 19–23
Moon, study of 85–6, 86
moral autonomy 64–5
motion 128–9

nationalism, avoidance of 39
natural selection see evolution of ideas; evolutionary theory
networks 142–4, 143
Newton, Isaac 29–30, 32–3
Newton's first law of mechanics 128–9
non-epistemic values 96–7, 99–100, 102
Norr, Roy 93
numeracy see statistics

objectivity in science 5, 84–7, 148, 152–4
omnicompetent citizens 12–14
Oreskes, N. 93–4
outsiders 113, 114–15

participatory democracy 12–13
paternalism 72–3
peer review 36–7
Peirce, Charles S. 34, 35
people 51, 55
philosophy
    role in public debate 134–45
    role in science 38
    of science 136–7
    see also history and philosophy, role of
Plato 34–5
pluralism 60–1, 63, 131
Poincaré, Henri 33

polarization 116
political change 132
politicians
   as experts 71
   relation with experts 13, 15, 21, 45–6
Popper, Karl 34, 40, 41, 56, 153
popularization of science 120
populism
   causes of 4–5
   and direct democracy 64–9
   slide towards 13
poverty 119
predictions 29
principles of democracy
   equality principle 47–51
   majority principle 51–5, 106, 109, 112
   separation of powers principle 55–8
prisoner dilemma 60
probabilistic correlation 139–40
probability
   and fake news 137–9
   in jury theorem 108, 110
   a priori probability 143
   statistical 95–6, 99–100
problem-solving 1, 44–5, 58–60
Program for International Student Assessment (PISA) 118
propaganda 13, 145
Prusiner, Stanley 93–4
pseudosciences 89–90
psychoanalysis 83–4
public engagement with science 113, 120
publication of scientific works 36–9, 122–3

quantum mechanics 38, 54, 91

reading skills 117–21
reconciliation problem (RP) 9–10, 13, 14
referenda 12
relativism 135, 146–52
relativity, general theory of 32, 32–3, 132
religious attitudes 116, 117
representative democracy
   characteristics of 3, 63
   comparison with direct democracy 11–12, 64, 65
   and competence 2

   as model of democracy 11
   politicians, role of 15
   and specialization of scientific knowledge (T_1) 1–2, 23, 45, 69, 71–2, 76–8, 105–6
technocracy, risks of 69–73
rights
   equality of 43–4, 48–9
   inequality of 44
Rousseau, Jean-Jacques, *Social Contract* 20, 64–5
Rovelli, Carlo 128
Russell, Bertrand 29, 84–5

Salmon, Wesley C. 140
science, how it works 3, 25–46, 52–4
   competition in 49–51
   consensus 3, 34, 36–7, 40, 42, 52, 83–4, 86
   criticism in a democracy 42–5, 54
   descriptive and normative aspects of science and democracy 25–6
   evolution of ideas 40–2
   fallibility of scientific knowledge 33–5
   inferences in science 28–30
   internet, role of 67–9
   methods of 17–18
   objectivity in science 84–7
   philosophy and 38
   philosophy of 18
   progress of theories 30–3, 52–3
   social character of science 16
   social control of science 35–40, 56
scientific academies 38–9
scientific change 30–3, 130–4
scientific concepts 16
scientific disinformation *see* disinformation and distrust
scientific ethics 92–7
scientific fraud 97–8
scientific literacy, need for (T_2) 6, 22–3, 105–24
   Condorcet's jury theorem 105, 106–13
   evaluation of competence 121–4
   experts, choice of 79
   illiteracy 115–21
   meaning of "scientific literacy" 113–15
scientific realism 85
scientists

evaluation of 121–4
inequality of 68
second thesis of the book *see* scientific literacy, need for
Semmelweiss, Ignaz 141
separation of powers principle 55–8
Shen, B. S. P. 113
smoking and health 92–4, 95–6, 139–40
Smolin, Lee 53
Snow, Charles 120
social anthropology 16–17
social cohesion 52
social contract 41–2
social inequalities 119
social media 68, 102, 121
social utility 50
sound 28
specialization of scientific knowledge (T_1) 1–2, 23, 45, 69, 71–2, 76–8, 105–5
spectators and agents 13
statistics 94–5, 114
see also probability
superstitious beliefs 90, 140–1
swindlers and charlatans 88–92
Switzerland 67

technocracy 2, 14, 45, 69–73
technology 2, 5, 67, 117
testing, medical 99–100
theses, of the book 1–2, 22–3
thirteen, unlucky 140–1
tobacco industry 92–4, 95–6
Tocqueville, Alexis de, *Democracy in America* 55
tolerance 35
trust
and doubts 92–7
in experts 45, 74–5
between politicians, citizens, and experts 15
in science 82–7
see also disinformation and distrust

understanding 114–15

vaccines 94, 102, 138–9
values
controllability of science 35–40
justifiability 44–5
in science 96–7, 98–102
scientific ethics 92–7
shared 21–2
see also autonomous choice; equality of rights
Vannoni, Davide 90
Voltaire 35

web
evaluation of competence 121–4
fake news 118
false theories spread on 102
science and 67–9
use in democracies 67
Weber, Max 11, 13, 96
Whewell, William 145
Whipple, M. 14
Wilks, Samuel S. 138

www.ingramcontent.com/pod-product-compliance
Lightning Source LLC
Chambersburg PA
CBHW061834300426
44115CB00013B/2383